Pathways to Environmental Sustainability

Roberta Salomone • Giuseppe Saija
Editors

Pathways to Environmental Sustainability

Methodologies and Experiences

Springer

Editors
Roberta Salomone
Department SEAM
University of Messina
Messina
Italy

Giuseppe Saija
Department SEAM
University of Messina
Messina
Italy

ISBN 978-3-319-03825-4 ISBN 978-3-319-03826-1 (eBook)
DOI 10.1007/978-3-319-03826-1
Springer Cham Heidelberg New York Dordrecht London

Library of Congress Control Number: 2013957488

Printed on acid-free paper

Springer is part of Springer Science+Business Media (www.springer.com)

Preface

This book puts together research experiences and solutions oriented to combine competitiveness and environmental sustainability, both on a methodological and applicative dimension, coming from academics, managers operating in various productive sectors and consultants. The aim of this publication is contributing to the growth and spread of scientific research and technological development for environmental sustainability, a topic whose importance is growing in Europe, involving various industries and professional contexts.

Pursuing this aim, the book gives an outline of the multiple application of the concept of environmental sustainability both providing an overview of various methodologies and tools useful in achieving sustainability targets and reporting on a very different experiences and case studies on sustainable management. Indeed, the basic idea that led the authors to edit this book was to bring together in a single volume:

- An overview of the multiple application of the concept of environmental sustainability, because the book contains examples of different methodologies and tools useful in pursuing environmental targets;
- Experiences and case studies applied in different productive sectors, embracing both industry experiences and research projects;
- Experiences and case studies applied in very different territorial context.

The development of this basic idea allowed the authors to edit a book that has a multidisciplinary European dimension, because, even if it mainly contains experiences and case studies related to the Italian context, also other countries practices are reported, furthermore authors come from different countries and activity sectors, and report on environmental issues of primary importance in Europe.

All the chapters of this book have been submitted to a double blind peer review and were organized in three main thematic parts:

- The first part covers some of the most interesting and emerging issues relating environmental sustainability methodologies, such as Industrial Ecology, Urban Metabolism, Life Cycle Assessment, Water Footprint and examples of some chemical technologies;

- The second part provides experiences and case studies of environmental sustainability applied in specific productive sectors, such as the electronic, the pharmaceutical, the energy, and the agri-food ones;
- The third part provides experiences and case studies of environmental sustainability specifically applied in territorial contexts on a local, regional or national scale, such as cities, industrial areas or countries.

This book is the result of research and dissemination activities realized with the support of:

The University of Messina

The Italian Academy of Commodity Science—AISME

R. Salomone
G. Saija

Contents

Part II Experiences and Case Studies in Productive Contexts

Part III Experiences and Case Studies in Territorial Contexts

Contributors

Ayse Guney Agar Department of Business Studies, Università di Roma Tre, Rome, Italy

Antonella Ambrosone Fabio Mataluni & C. srl, Montesarchio, BN, Italy

Gabriella Arcese Department of Business Studies, Università di Roma Tre, Rome, Italy

Dina Margrethe Aspen Department of Industrial Economics and Technology Management, Norwegian University of Science and Technology, NTNU, Trondheim, Norway

Bruno Andrea Autelitano University of Udine, Udine, Italy

Grazia Barberio Environmental Technologies Technical Unit, ENEA, Roma, Italy

Jan C. J. Bart Department of Electronics Engineering, Chemistry and Industrial Engineering (DIECII), University of Messina, Messina, Italy

Antonio Battiato ST Microelectronics—site of Catania, Catania, Italy

Paolo Bogoni "Bruno de Finetti" Dept. of Economics, Business, Mathematics and Statistics, University of Trieste, Trieste, Italy

Claudia Brunori Environmental Technologies Technical Unit, ENEA, Roma, Italy

Luigi Bruzzi Department of Physics, University of Bologna, Bologna, Italy

Grazia Calabrò Department SEAM, University of Messina, Messina, Italy

Barbara Campisi "Bruno de Finetti" Dept. of Economics, Business, Mathematics and Statistics, University of Trieste, Trieste, Italy

Francesco Cappello Energy Counseling Centre Sicilia, ENEA, Palermo, Italy

Luciano Ceccon Department of Economics and Statistics, University of Udine, Udine, Italy

Maurizio Cellura Dipartimento di Energia, Ingegneria dell'Informazione e Modelli Matematici, University of Palermo, Palermo, Italy

Salvatore Chiricosta Department SEAM, University of Messina, Messina, Italy

Giuseppe Cirillo Department of Pharmacy, Health and Nutritional Sciences, University of Calabria, Rende, Italy

Stefano Cucurachi Institute of Environmental Sciences, University of Leiden, Leiden, The Netherlands

Laura Cutaia Environmental Technologies Technical Unit, ENEA, Roma, Italy

Isabella D'Antuoni CRIOL, Research Centre for Olive Oil Industry–Industria Olearia Biagio Mataluni srl, Montesarchio, BN, Italy

Pierpaolo Dell'Omo DIAEE, Sapienza University of Rome, Rome, Italy

Pauline Deutz Department Geography, Environment and Earth Sciences, University of Hull, Hull, UK

Salvatore Falco CRIOL, Research Centre for Olive Oil Industry–Industria Olearia Biagio Mataluni srl, Montesarchio, BN, Italy

Annik Magerholm Fet Department of Industrial Economics and Technology Management, Norwegian University of Science and Technology, NTNU, Trondheim, Norway

Mario Fontana Dipartimento di Energia, Ingegneria dell'Informazione e Modelli Matematici, University of Palermo, Palermo, Italy

Maria Francesca Milazzo Department of Electronics Engineering, Chemistry and Industrial Engineering (DIECII), University of Messina, Messina, Italy

Elisa Gatto Department of Economics, Business, Environment and Quantitative Methods, University of Messina, Messina, Italy

Paola Geatti Department of Chemistry, Physics and Environment, University of Udine, Udine, Italy

Reinout Heijungs Department of Econometrics and Operations Research, VU University Amsterdam, Amsterdam, The Netherlands

Institute of Environmental Sciences, University of Leiden, Leiden, The Netherlands

Giuseppe Ioppolo Department SEAM, University of Messina, Messina, Italy

René Kleijn Institute of Environmental Sciences, University of Leiden, Leiden, The Netherlands

Joanna Kulczycka AGH University of Science and Technology, Cracow, Poland

Francesco Lanuzza Department of Economics, Business, Environment and Quantitative Methods, University of Messina, Messina, Italy

Lukasz Lelek Mineral and Energy Economy Research Institute of The Polish Academy of Sciences, Cracow, Poland

Mariarosaria Lombardi Department of Economics, University of Foggia, Foggia, Italy

Sonia Longo Dipartimento di Energia, Ingegneria dell'Informazione e Modelli Matematici, University of Palermo, Palermo, Italy

Maria Claudia Lucchetti Department of Business Studies, Università di Roma Tre, Rome, Italy

Francesca Luciani CRIVIB, Istituto Superiore di Sanità, Rome, Italy

Marina Magerøy Department of Industrial Economics and Technology Management, Norwegian University of Science and Technology, NTNU, Trondheim, Norway

Dagfinn Malnes The Norwegian EPD Foundation, Oslo, Norway

Erika Mancuso Environmental Technologies Technical Unit, ENEA, Roma, Italy

Alessandro Manzardo Department of Industrial Engineering, Quality and Environment Research Center, University of Padova, Padova, Italy

Elena De Marco CRIOL, Research Centre for Olive Oil Industry–Industria Olearia Biagio Mataluni srl, Montesarchio, BN, Italy

Federica Marinatto University of Trieste, Trieste, Italy

Alba Marino Department of Economics, Business, Environment and Quantitative Methods, University of Messina, Messina, Italy

Anna Mazzi Department of Industrial Engineering, Quality and Environment Research Center, University of Padova, Padova, Italy

Daniele Milone Dipartimento di Energia, Ingegneria dell'Informazione e Modelli Matematici, University of Palermo, Palermo, Italy

Antonio Mione Environmental Technologies Technical Unit, ENEA, Pisa, Italy

Marina Mistretta Dipartimento Patrimonio, Architettura, Urbanistica, University of Reggio Calabria, Reggio Calabria, Italy

Fabio Mondello Department of Economics, Business, Environment and Quantitative Methods, University of Messina, Messina, Italy

Roberto Morabito Environmental Technologies Technical Unit, ENEA, Roma, Italy

Raffaele Mosca Department of Economic Studies, G. d'Annunzio University, Pescara, Italy

Camillo Mungiguerra Technical Unit for Information Systems and ICT Development, ENEA, Roma, Italy

Bruno Notarnicola Ionian Department of Law, Economics and Environment, University of Bari "Aldo Moro", Taranto, Italy

Veronica Novelli Department of Economics and Statistics, University of Udine, Udine, Italy

Ortensia I. Parisi Department of Pharmacy, Health and Nutritional Sciences, University of Calabria, Rende, Italy

Pasquale Pazienza Department of Economics, University of Foggia, Foggia, Italy

Luigia Petti DEC—Department of Economic Studies, G. D'Annunzio University, Pescara, Italy

Nevio Picci Department of Pharmacy, Health and Nutritional Sciences, University of Calabria, Rende, Italy

Raffaella Preti Department of Management, Sapienza University of Rome, Rome, Italy

Francesco Puoci Department of Pharmacy, Health and Nutritional Sciences, University of Calabria, Rende, Italy

Andrea Raggi Department of Economic Studies, G. d'Annunzio University, Pescara, Italy

Paola Karina Sanchez Ramirez DEC—Department of Economic Studies, G. D'Annunzio University, Pescara, Italy

Roberto Rana Department of Economics, University of Foggia, Foggia, Italy

Pietro Alexander Renzulli Ionian Department of Law, Economics and Environment, University of Bari "Aldo Moro", Taranto, Italy

Donatella Restuccia Department of Pharmacy, Health and Nutritional Sciences, University of Calabria, Rende, Italy

Ornella Li Rosi Technical Unit for Advanced Technologies for Energy and Industry, ENEA, Ispra, Varese, Italy

Simona Saccà Department SEAM, University of Messina, Messina, Italy

Giuseppe Saija Department SEAM, University of Messina, Messina, Italy

Roberta Salomone Department SEAM, University of Messina, Messina, Italy

Maria Savarese Fabio Mataluni & C. srl, Montesarchio, BN, Italy

Antonio Scipioni Department of Industrial Engineering, Quality and Environment Research Center, University of Padova, Padova, Italy

Francisco Serrano-Bernardo Department of Civil Engineering, University of Granada, Granada, Spain

Guido Signorino Department of Economics, Business, Environment and Quantitative Methods, University of Messina, Messina, Italy

Pasquale Spezzano Environmental Technologies Technical Unit, ENEA, Roma, Italy

Francesco Spina Department of Electronics Engineering, Chemistry and Industrial Engineering (DIECII), University of Messina, Messina, Italy

U. Gianfranco Spizzirri Department of Pharmacy, Health and Nutritional Sciences, University of Calabria, Rende, Italy

Valentino Tascione Department of Economic Studies, G. d'Annunzio University, Pescara, Italy

Giuseppe Tassielli Ionian Department of Law, Economics and Environment, University of Bari "Aldo Moro", Taranto, Italy

Alessio Tola Department of Economic Sciences, University of Sassari, Sassari, Italy

Enrique Toscano Bologna, Italy

Caterina Tricase Department of Economics, University of Foggia, Foggia, Italy

Maria Marcella Tripodo Department of Chemistry Sciences, University of Messina, Messina, Italy

Cassia Maria Lie Ugaya PPGEM—Post Graduation Program of Materials and Mechanical Engineering, Federal Technological University of Parana, Curitiba, Brazil

Simona Verità Bologna, Italy

Giuliana Vinci Department of Management, Sapienza University of Rome, Rome, Italy

List of Figures

List of Tables

About the Editors

Roberta Salomone is Associate Professor of Environmental Management and Industrial Ecology at the Department of Economics, Business, Environment and Quantitative Methods—University of Messina; her main research topics are industrial ecology, quality and environmental management systems, Life Cycle Assessment and other environmental assessment tools, including the study of potential integration among tools and systems of environmental management. She is member of the Steering Committee of the Italian Academy of Commodity Science and of the Steering Committee of the Italian LCA Network.

Giuseppe Saija is Full Professor of Quality and Environmental Management at the Department of Economics, Business, Environment and Quantitative Methods, University of Messina; his main research topics are quality and environmental management systems, integrated management systems (quality, environment, safety, ethics, etc.), food quality.

Part I
Methodologies and Tools for Environmental Sustainability

Chapter 1
Food for Thought: Seeking the Essence of Industrial Symbiosis

Pauline Deutz

Abstract Researchers and practitioners would benefit from a definition of industrial symbiosis which clearly distinguishes essential from contingent characteristics. The definition also needs to be translatable between both language and policy contexts. Industrial symbiosis is herein defined as a flow of underutilized resource(s) (comprising substances and/or objects and/or energy), from an entity which would otherwise discard them, to another entity which uses them as a substitute for new resources. Choice of terms is justified by reference to academic and policy literature. This definition has an underlying assumption of resource efficiency, by contrast to other approaches which mistakenly emphasized economic benefits, which are contingent rather than essential characteristics.

Keywords Industrial symbiosis · Industrial ecology · Policy · Resources · Waste

1.1 Introduction

Industrial symbiosis (IS) is a field of study and environmental business practice within the interdisciplinary field of industrial ecology (IE). Core to the idea of IS is a collaborative approach to the extraction of value from otherwise underexploited resources. However, after two decades of debate, the definition of IS remains contested (Lombardi and Laybourn 2012), even whilst authors develop theorizations (e.g., Paquin and Howard-Grenville 2012). Without a clear understanding of what is IS, and the implications of that, efforts to build theories may be at cross purposes and possibly emphasize incidental rather than key characteristics. A clear understanding of underlying assumptions and their implications for methodology is critical to any research endeavor (Creswell 2003). Clarity and consistency of definition are no less a concern for policy initiatives, which for IS are found across the globe (Lombardi and Laybourn 2012). Assumptions underlying research relate to ontology and epistemology, which respectively ask questions about what exists and how we can gain knowledge of it (Spash 2012). It is beyond the scope of this chapter

P. Deutz (✉)
Department Geography, Environment and Earth Sciences,
University of Hull, Hull HU6 7RX, UK
e-mail: p.deutz@hull.ac.uk

R. Salomone, G. Saija (eds.), *Pathways to Environmental Sustainability*,
DOI 10.1007/978-3-319-03826-1_1,

to address the ontology and epistemology of IS comprehensively. I also cannot address in detail the variety of IS policy initiatives, rather the aim is to consider the underlying commonalities. The chapter focuses on the urgent and critical ontological question of what is IS.

Although studied by engineers and natural scientists as well as social scientists, IS is a social phenomenon (Andrews 2000/2001). Arguably, this makes IS more complex to define than either the physical objects that more typically comprise the object of natural science study (Sayer 1984), or the engineered systems that are intentionally defined and controlled to suit a specific purpose (Andrews 2000/2001). For social phenomena the boundaries between object and ideas about the object can be hard to disentangle, with ideas (and potentially the object) heavily contingent on historical and geographic context (Sayer 1984); the "object" of social science study may be an idea (Sayer 1984). Indeed, in some usages IS is an idea, i.e., that industry could and should be reorganized into industrial ecosystems (see below). This chapter argues, however, that the term industrial ecosystem should be reserved for the idea, or ideal, of that reorganization. IS could better be defined as a distinct, observable phenomenon.

This chapter argues that an unambiguous definition of IS is required, which identifies essential features, as distinct from contingent properties. The latter may accurately describe some instances of IS in practice, but should not be considered defining characteristics. Furthermore, any definition of IS needs to be translatable between different language and policy contexts. The proposed definition identifies resource efficiencies as the essence of IS, justifying the choice of terms by reference to the IS literature and relevant policy documents.

1.2 Industrial Symbiosis and Industrial Ecosystems

Almost from the beginning of the widespread usage of the term IS there have been both different usages of the term, and other expressions used synonymously. Current academic and policy interest in IS (and IE more generally) stems from the fortuitous near coincidence of the Frosch and Gallopoulos (1989) article advocating "industrial ecosystems" and the 1990 discovery of an apparent industrial ecosystem in practice at Kalundborg, Denmark (e.g., Jacobsen 2006). The *Financial Times* article that publicized Kalundborg called its network of waste exchanges an "industrial ecosystem", presenting IS as "the term preferred by the locals" (Knight 1990, p. 15). An industrial ecosystem applies the lessons that industrial ecology draws from biological systems (Frosch and Gallopoulos 1989; Korhonen and Baumgartner 2009). The term Industrial ecosystem continues to appear in the IS literature to evoke a resource efficiency network of the type displayed at Kalundborg (e.g., Chertow and Ehrenfeld 2012).

IS has taken root in the literature in a sense similar to that of an industrial ecosystem (e.g., Wolf et al. 2007; Lombardi et al. 2012). The most widely cited definition of IS (Lombardi and Laybourn 2012) comes from Chertow (2000): "The part of industrial ecology known as industrial symbiosis engages traditionally separate industries in a collective approach to competitive advantage involving physical exchange of materials, energy, water and by-products. The keys to industrial symbiosis are collaboration and the synergistic possibilities offered by geographic proximity."

This definition closely corresponds to circumstances at Kalundborg, and could be seen as an elucidation of the idea of an industrial ecosystem, rather than a separate empirical phenomenon.

However, there have been, and still are, alternative approaches, which see symbiosis as the individual relationships, the sum of which may comprise an industrial ecosystem: "Although there are many examples of symbiosis involving the exchange of material and the cascading of energy and water … multifaceted industrial ecosystems are few and far between." (Côte and Cohen-Rosenthal 1998; see also Behera et al. 2012). The centrality of inter-firm resource transactions is important to many authors' concepts of IS (e.g., Jensen et al. 2011). The focus on individual inter-firm relationships corresponds to the perspectives of participating firms, even when they happen to be part of a network of such relationships (Posch 2010). In addition, however, terms such as "synergies" are used for the individual inter-firm relationships (e.g., Mangan 1997; Jensen et al. 2011). Potential ontological confusion is compounded by the use of expressions such as "regional synergies" (van Beers et al. 2007) and "by-product synergy networks" (Cimren et al. 2011) as equivalents to industrial ecosystems. The individual resource flow relationship is the foundation for the definition of IS proposed herein; it offers a distinct concept and highlights the essential IS property of resource conservation potential.

1.3 (Re)Defining Industrial Symbiosis

The following definition of IS is proposed:

> Industrial symbiosis is a flow of underutilized resource(s) (comprising substances and/or objects and/or energy), from an entity which would otherwise discard them, to another entity which uses them as a substitute for new resources.

All work on IS is clear that IS does not refer to a "business as usual" transaction between bodies, i.e., the sale of a good for the purpose for which it was intended. A defining characteristic of IS is the idea of extracting additional value from a resource that might otherwise be disposed of (or dispersed to the atmosphere or as effluent to a waterway). The definition does not preclude a potential requirement for resource processing prior to use by the receiving entity (e.g., as described by Behera et al. 2012 at Ulsan in South Korea). However, the receiving entity is not primarily a waste disposal facility (Chertow 2007); IS is an alternative to disposal. The term "entity" is used (Chertow 2007) to avoid being prescriptive on the type(s) of organization that may be involved. Critically, some level of separation between entities is implied; system rather than organizational boundaries distinguish IS (and IE) from other environmental management approaches (Deutz 2009).

IS transactions can, as at Kalundborg, include substances in solid, liquid, or gaseous form and/or energy. I suggest that IS is defined to include all these types of "resources", as the principle of a transaction that saves resources is more fundamental to the phenomenon than the implications of differences between those resources. This is not to preclude the potential for differences between resource types to have implications for IS (c.f., van Beers et al. 2007; Ashton 2008), rather to confirm IS could involve any of them. "Underutilised" (Jensen et al. 2011) emphasizes that the resources have been consumed in some sense. A related key point is that the originating entity does not want the resource. Thus, IS transaction could (*in extremis*) include raw materials that have been ordered in excess by a manufacturer, and would otherwise be thrown away, but not "business as usual" sale of raw materials by an extractive firm. IS could also be a means for a manufacturer to extract some value from "business as usual" products for which there is no "business as usual" demand. Significantly, I use the word flow rather than exchange, to avoid any suggestion that there needs to be a two-way trade of materials between any two entities for IS to be observed. Notably, however, any flow has to be accommodated in the institutions of exchange that govern inter-firm relationships in general (Miller and Ford 2007). The return flow within an IS relationship could be material, but also financial, a service or potentially based on the assumption of a future return. IS flows do often literally flow in that they comprise "continuous-process waste streams" (Lyons 2007), but this is an observation that merits more investigation, not a necessary characteristic of IS.

1.4 Language and Policy Context

The proposed definition has deliberately avoided terms such as "waste", "by-product" and "residual" which are commonly used in IS to refer to the underutilized resources being transferred. Sometimes a distinction between these terms is clearly implied (e.g., Chertow and Ehrenfeld 2012), though not necessarily elucidated. In other work two or more of the terms are used interchangeably (e.g., Mangan 1997; Posch 2010). Readers therefore will inevitably be making their own interpretations according to what is common and/or policy usage (not necessarily the same thing) in their geographic context. Significantly, waste, by-product and residual all have distinct (but not mutually exclusive) dictionary definitions in the English language (Table 1.1). Equivalents do not necessarily exist in other languages.

Furthermore, these waste-related terms have specific policy connotations, which vary at scale of governance (e.g., UN usage differs from EU; Table 1.1), are likely to vary between political jurisdictions and are vulnerable to political redefinition over time. The UN usage, for example, implies that waste is a general term for something that is to be discarded; if it has value can be called a product (sic), or is otherwise a residual (Table 1.1). Conversely, the EU uses residual as a term for something that is waste if it has no assured route to reuse, or a by-product if it meets all of several conditions (Table 1.1). In addition, the same substance can at different times,

Table 1.1 Comparison of definitions of terms frequently used in discussions, descriptions and definitions of IS. Quotation marks indicate exact quotes; other expressions have been paraphrased for brevity

Term	Dictionary definition (Chambers Dictionary 2013)	European Union definition (EU 2008)	United Nations definition (UN 2012)
Waste	Something no longer needed in its present form which must be processed; or refuse, rubbish	"any substance or object which the holder discards or intends or is required to discard" (EU 2008, p. 9)	"Discarded materials no longer required by the owner or user" (UN 2012, p. 51)
By-product	"A secondary and often commercially important product that is formed at the same time as the main product during a chemical reaction or manufacturing process"	"A substance or object, resulting from a production process, the primary aim of which is not the production of that item" is a by-product if all the following conditions hold: • further use is certain, and legal • the substance is produced as "integral part of a production process," and • requires no preprocessing "beyond normal industrial practice" (EU 2008, p. 11)	Not applicable: the UN uses the word "product" in a context that approximates the use of by-product by the other sources: Waste with a positive value "is considered a product…rather than a residual" (UN 2012, p. 51), i.e., discarded material exchanged between economic units, for example scrap metal, for which the discarder receives payment
Residual	"Something that remains left over as a residue", e.g., left over after a process (e.g., evaporation) or when other parts have been taken away	'Not used. In UK 'residue' covers both waste and by-products (DEFRA 2012)'	"Flows of solid, liquid and gaseous materials, and energy that are discarded, discharged… through processes of production, consumption or accumulation" (UN 2012, p. 49)

depending on demand and supply, technological developments and social expectations, as well as regulatory requirements, be a challenge to dispose of, a lucrative by-product, or even become product in its own right (Desrochers 2009). These elements will vary geographically, too. The proposed definition of IS, therefore, does not discriminate between transactions according to their assumed ease of finding a receiver, though one might classify different types of IS according to such criteria.

Avoiding specific terms such as waste and by-product helps to define IS in a manner accessible to different languages and policy contexts. The proposed definition is a deliberate attempt to be broad. It should not exclude very much that currently exists under the banner of IS. But setting an inclusive definition is not to stifle debate. On the contrary, this should open a lively debate on the observed forms of IS, the circumstances that favor them and potential theorizations, without sensitivity as to what is or is not IS. However, selecting the resource based relationship as the essence of IS is consciously relegating other aspects from the Chertow (2000) and related definitions to the status of contingencies.

1.5 Contingent Characteristics of IS

Several characteristics of IS that are commonly included in definitions are excluded from the one proposed here on the grounds that they are not essential to the definition. Rather, as will be briefly discussed in this section, they are contingent characteristics that can take on a range of different forms or values, but should not be seen as determining whether or not a given phenomenon is IS.

The environmental benefits of IS are often assumed rather than measured (Boons et al. 2011), albeit the Kalundborg industrial ecosystem and others do have impressive resource conservation statistics (e.g., Jacobsen 2006; Behera et al. 2012). Chertow and Ehrenfeld (2012) contend that an IS transaction should be environmentally beneficial by definition, i.e., any non-environmentally sound underutilized resource transaction would not be deemed IS. However, this is problematic. Detailed and precise environmental impact assessments are difficult to undertake, even if assumptions can be made about what is and is not harmful to the environment. It does not seem desirable to make a property, which is likely unknown, and potentially unknowable, a defining characteristic of IS. This is not to say that IS should be promoted for its own sake, regardless of environmental side effects. Ideally the latter should be at least estimated on a case by case basis. An existing relationship found to be of doubtful environmental benefit, would bear closer investigation, but would not cease to be IS.

Economic benefits, or competitive advantage, are a widely cited aspect of IS (e.g., Chertow 2000; Cimren et al. 2011; Lombardi and Laybourn 2012). Economic savings not only can occur from IS, but are seen as critical to IS agreements (e.g., Jacobsen 2006). However, observing the coincidence of economic and environmental benefit has resulted in undue confidence that they are associated with each other. In the US context, initiatives to construct industrial ecosystems were far more often led by economic development than environmental protection agencies (Deutz and Gibbs 2008). Furthermore, emphasizing competitive advantages of IS distracts from the arguably more important point of the potential resource efficiencies. Defining IS by the essential characteristic of resource flow helps to clarify a likely more transferable lesson from Kalundborg than the vision of a collaborative ecosystem. Economic benefits, or disbenefits, are highly contingent, e.g., on the price of raw materials, the cost of waste disposal. If resource conservation were a sufficient social/political priority, IS could be a regulatory requirement for one or both entities involved, irrespective of the financial implications. These may seem rather radical and potentially unwise suggestions. However, the principle to establish is simply that IS is in essence a resource conservation rather than necessarily an economic efficiency tool. It is a political decision as to how far environmental protection should be limited to initiatives that may provide cost savings. Social benefits from IS are scarcely mentioned in the literature, and as with economic, are potential empirical features of a given IS, highly worthy of exploration, but not a defining characteristic.

In addition, the proposed definition says nothing about the geographic distance between the entities or networking. The local scale is often prioritized in IS projects (e.g., Wolf et al. 2007; Van Berkel et al. 2009), but it can be found in practice at scales up to and including global (Lyons 2007). Networking, and proximity may, possibly in combination, be important precursors to IS (Deutz and Gibbs 2008; Jensen et al. 2011). However, they are not defining characteristics, and neither necessary nor sufficient to engender IS. In keeping with Chertow's reference to "physical exchange", I am excluding inter-firm networking and knowledge transfer from the definition of IS (in contrast to Lombardi and Laybourn 2012); "industrial ecology" can be used for such nonphysical environmental cooperation between entities.

Thus several previously key aspects of IS are shown to be contingent properties, influenced by wide range of potential circumstances. The final section considers the implications of confining the definition of IS to the essential characteristics.

1.6 Conclusions: Changing the Assumptions of IS

This chapter has proposed a definition of industrial symbiosis which identifies the resource conservation relationships between two or more entities as the essential characteristic of the phenomenon. The resulting definition is therefore specific (there must be a physical flow of the kind specified), whilst encompassing a wide range of possibilities (e.g., in terms of geographic scale, number of participants, whether or not the flows are economically attractive to any or all participants). It is also designed to be transferable across language and policy contexts.

IS under the proposed definition becomes essentially an option for environmental protection. There is an underlying assumption that the environment should be protected, but not that IS is necessarily the most beneficial approach in any given circumstance. Removing the normative assumptions often surrounding IS (e.g., Boons and Roome 2001), may greatly ease communications between the different epistemologies contributing to IS research. Calls for social science contributions to what had been an engineering field (Vermeulen 2006), have been heeded (e.g., Doménech and Davies 2010). However, communications between the various epistemologies are limited. Clarifying the ontology of IS is at least an important first step to facilitate communication across the field.

In policy terms, IS as a resource efficiency measure offers potentially a strong tool for environmental protection, whereas it has proved a weak tool for economic development (Deutz and Gibbs 2008). The drivers for IS could be enhanced by recycled materials requirements for manufacturers, for example (the EU Producer Responsibility Regulations provide a gentle signal; Deutz 2009). However, the proposed definition by no means mandates a regulatory approach to IS, or precludes voluntary IS relationships where there happens to be financial benefit.

The proposed definition of IS is designed to be a prelude to much further research, whether primarily empirical or theoretical. Research is needed to identify, describe, classify, understand, measure/model the environmental impacts of, and ultimately to build theories to attempt to explain, all the many spatially and temporally contingent variants of IS and how stakeholders engage with them. This work would feed directly into policy considerations. What initiative would work best in a given circumstance? How would firms react? Opinions can of course differ on the appropriateness of IS under given circumstances, or its social/environmental value in general. Such debates are extremely healthy, but easier if not involving territorial disputes over what is and is not IS.

References

Andrews CJ (2000/2001) Restoring legitimacy to the systems approach. IEEE Technology and Society Magazine. Winter 2000/2001:38–44

Ashton W (2008) Understanding the organization of industrial ecosystems: a social network approach. J Ind Ecol 12:34–51

Behera SK, Kim J-H, Lee S-Y Suh S, Park H-S (2012) Evolution of 'designed' industrial symbiosis networks in the Ulsan eco-industrial park: 'research and development into business' as the enabling framework. J Clean Production 29-30:103–112

Boons F, Roome N (2001) Industrial ecology as a cultural phenomenon: on objectivity as a normative position. J Ind Ecol 4:49–54

Boons F, Spekkink W, Mouzakitis Y (2011) The dynamics of industrial symbiosis: a proposal for a conceptual framework based upon a comprehensive literature review. J Clean Production 19:905–911

Chambers Dictionary (2013) Chambers online 21st century dictionary. http://www.chambers.co.uk/. Accessed 13 Jan 2013

Chertow MR (2000) Industrial symbiosis: literature and taxonomy. Annu Rev Energy Environ 25:313–337

Chertow MR (2007) "Uncovering" industrial symbiosis. J Ind Ecol 11:11–30

Chertow MR, Ehrenfeld J (2012) Organizing self-organizing systems: toward a theory of industrial symbiosis. J Ind Ecol 16:13–27

Cimren E, Fiksel J, Posner ME, Sikdar S (2011) Material flow optimization in by-product synergy networks. J Ind Ecol 15:315–332

Côte RP, Cohen-Rosenthal E (1998) Designing eco-industrial parks: a synthesis of some experiences. J Clean Production 6:181–188

Creswell JW (2003) Research design: qualitative. Quantitative and mixed methods approaches, 2nd edn. Sage, Thousand Oaks

DEFRA, 2012, *Guidance on the legal definition of waste and its application.* DEFRA, London. https://www.gov.uk/government/uploads/system/uploads/attachment_data/file/69590/pb13813-waste-legal-def-guide.pdf 20/12/13

Desrochers P (2009) Does the invisible hand have a green thumb? Incentives, linkages, and the creation of wealth out of industrial waste in Victorian England. Geogr J 175:3–16

Deutz P (2009) Producer responsibility in a sustainable development context: ecological modernisation or industrial ecology? Geogr J 175:274–285

Deutz P, Gibbs D (2008) Industrial ecology and regional development: eco-industrial development as cluster policy. Reg Stud 42:1313–1328
Doménech T, Davies M (2010) The role of embeddedness in industrial symbiosis networks: phases in the evolution of industrial symbiosis networks. Bus Strategy Environ 20:281–296
EU (2008) Waste Framework Directive. 2008/98/EC http://eur-lex.europa.eu/LexUriServ/LexUriServ.do?uri=OJ:L:2008:312:0003:0030:EN:PDF. Accessed 13 Jan 2013
Frosch RA, Gallopoulos NE (1989) Strategies for manufacturing. Sci Am 261:94–102
Jacobsen NB (2006) Industrial symbiosis in Kalundborg, Denmark: a quantitative assessment of economic and environmental aspects. J Ind Ecol 10:239–255
Jensen PD, Basson L, Hellawell EE, Bailey MR, Leach M (2011) Quantifying 'geographic proximity': experiences from the United Kingdom's National Industrial Symbiosis Programme. Resour Conserv Recycl 55:703–712
Knight P (1990) A rebirth of the pioneering spirit. Financial Times 14.11.1990, p. 15
Korhonen J, Baumgartner RJ (2009) The industrial ecosystem balanced scorecard. Int J Innov Sustain Dev 4:24–42
Lombardi DR, Laybourn P (2012) Redefining industrial symbiosis: crossing academic-practitioner boundaries. J Ind Ecol 16:28–37
Lombardi DR, Lyons D, Shi H, Agarwal A (2012) Industrial symbiosis: testing the boundaries and advancing knowledge. J Ind Ecol 16:2–7
Lyons DI (2007) A spatial analysis of loop closing among recycling, remanufacturing and waste treatment firms in Texas. J Ind Ecol 11:43–54
Mangan AT (1997) By-product synergy: a strategy for sustainable development. University of Texas. http://repositories.lib.utexas.edu/bitstream/handle/2152/15226/tbr-1997.08-mangan.pdf?sequence=2. Accessed 13 Jan 2013
Miller VV, Ford MW (2007) Exchanges: physical, economic, but also institutional. Prog Ind Ecol 4:288–305
Paquin RL, Howard-Grenville J (2012) The evolution of facilitated industrial symbiosis. J Ind Ecol 16:83–93
Posch A (2010) Industrial recycling networks as starting points for broader sustainability-oriented cooperation? J Ind Ecol 14:242–257
Sayer A (1984) Method in social science: a realist approach. Hutchinson University Library, London
Spash CL (2012) New foundations for ecological economics. Ecol Econ 77:36–47
UN (2012) System of environmental-economic accounting central framework. White Cover publication. http://unstats.un.org/unsd/envaccounting/White_cover.pdf. Accessed 13 Jan 13
Van Beers D, Corder G, Bossilkov A, van Berkel R (2007) Industrial symbiosis in the Australian minerals industry. J Ind Ecol 11:55–72
Van Berkel R, Fujita T, Hasimoto S, Fjuii M (2009) Quantitative assessment of urban and industrial symbiosis in Kawasaki Japan. Environ Sci Technol 43:1271–1281
Vermeulen WJV (2006) The social dimensions of industrial ecology: on the implications of the inherent nature of social phenomena. Prog Ind Ecol 3:574–598
Wolf A, Eklund M, Söderström M (2007) Developing integration in a local industrial ecosystem - an explorative approach. Bus Strat Environ 16:442–455

Chapter 2
LCA and Linear Programming for the Environmental Optimization of Waste Management Systems: A Simulation

Valentino Tascione, Raffaele Mosca and Andrea Raggi

Abstract Life Cycle Assessment (LCA), as a tool for assessing environmental performance, can also be implemented in order to help decision makers choose the best alternatives of integrated waste management systems (IWMS) in terms of environmental impacts. In order to avoid making assumptions, linear programming (LP) models integrated to LCA may allow us to identify an optimum scenario of IWMS from the environmental point of view, among all those theoretically available. We developed an LP model and a multi-objective LP (MLP) model that aim to identify the optimal allocation of waste to minimize environmental impacts. In this research the developed models were tested through a simulation based on realistic data to verify their validity.

Keywords Waste management system · Waste management scenario · Linear programming model · Life cycle assessment · Optimization model

2.1 Introduction

Life Cycle Assessment (LCA) methodology can be also applied to assess the environmental performance of an integrated waste management system (IWMS) or to identify the system with the best performance through a comparative analysis of different scenarios. This methodology can also take into consideration environmental benefits related to certain options, such as recycling (Fukushima and Hirao 2002).

In the planning phase of IWMS, we can build scenarios to be compared. We define an integrated waste management scenario as a combination between the types of waste produced and the types of disposal and treatment processes that are technically and economically available. As already shown in previous work (Tascione and Raggi 2010), the choice of scenarios (or alternatives) to be compared in an LCA study applied to the IWMS is typically based on various criteria, that are more or less subjective. Although the scenarios may be defined in a valid manner according to the criteria identified in the literature (Tascione and Raggi 2010), the number of

A. Raggi (✉) · V. Tascione · R. Mosca
Department of Economic Studies, G. d'Annunzio University, 65127 Pescara, Italy
e-mail: a.raggi@unich.it

R. Salomone, G. Saija (eds.), *Pathways to Environmental Sustainability*,
DOI 10.1007/978-3-319-03826-1_2,

scenarios compared is often limited, due to the lack of time and resources available for studies. This means decision makers are not able to identify the environmentally optimum scenario, but instead must identify the one that is environmentally preferable within a limited and default number of alternatives and hypotheses. Two models based on linear programming (LP) were developed which try to assist decision makers in identifying an IWM scenario as close as possible to the optimum system in terms of environmental performances, with the integration of LCA impact factors concerning waste processing and transport (Mercuri and Raggi 2004; Tascione and Raggi 2012). The first model considers just one impact category at a time (e.g., Global Warming, Acidification, etc.) and identifies the optimum solutions which minimize each specific impact. The second model minimizes more than one impact category at the same time; for this reason it is referred to as a multi-objective LP (MLP) model.

2.2 The Problem with a Single Impact: LP Model

The parameters of the problem, defined below as problem *P*, are:

- $N = \{1, \ldots, n\}$ = set of waste fractions (e.g., glass, aluminium, …);
- $M = \{1, \ldots, m\}$ = set of destinations, i.e., types of waste treatment/disposal (e.g., incineration, recycling, …);
- N_i = unit(s) of waste *i* to allocate/distribute to the destinations, for $i = 1, \ldots, n$, in a time interval *T*;
- M_j = maximum unit(s) of waste that the destination *j* can accept, $j = 1, \ldots, m$, in a time interval *T*;
- b_{ij} = unit(s) of environmental impact to allocate 1 unit of waste *i* to the destination *j*, for $i = 1, \ldots, n, j = 1, \ldots, m$;
- d_j = distance from the waste collection centre (a single site) to the place of destination *j*;
- γ = unit(s) of environmental impact to transfer 1 unit of waste over 1 unit of distance.

Problem *P*: determine an allocation/distribution of the total amount of each waste fraction to the destinations, in order to minimize a single environmental impact.

We introduce the following decision variables:

x_{ij} = unit(s) of waste *i* allocated to the destination *j*, for $i = 1, \ldots, n,\ j = 1, \ldots, m$.

$\sum_{i=1}^{n} N_i \leq \sum_{j=1}^{m} M_j$ can be assumed. If this does not happen then you can always be redirected to this assumption by adding a residual destination consisting of the non-treatment of waste or another residual destination, associated with its environmental impact.

The model is the following: min $\sum_{i=1}^{n}\sum_{j=1}^{m}(\gamma d_j + b_{ij})x_{ij}$ subject to:

$$\sum_{j=1}^{m} x_{ij} = N_i \text{ for } i = 1, \ldots, \text{n}; \sum_{i=1}^{n} x_{ij} \leq M_j \text{ for } j = 1, \ldots, m;$$

$$\text{x}_{ij} \geq 0 \text{ for } i = 1, \ldots, n, \text{ for } j = 1, \ldots, m.$$

As stated previously, this model permits the solution of a problem related to a single environmental impact; however, it may be useful considering different impacts at the same time.

2.3 The Problem with Multiple Impacts: MLP Methods

We assume that there is a set $L = \{1, \ldots, l\}$ of environmental impacts and the following parameters are associated to the environmental impact *k*, for $k=1, \ldots, l$: b_{ijk} (units of environmental impact *k* to allocate 1 unit of the material *i* to the destination *j*, for $i=1, \ldots, n$, for $j=1, \ldots, m$); γ_k (units of environmental impact k to transfer 1 unit of material over 1 unit of distance).

For convenience we can write:

$$X = \left\{x = \left\{x_{ij}\right\}, \text{for } i = 1, \ldots, n, \text{ for } j = 1, \ldots, m{:}\sum_{j=1}^{m} x_{ij} = N_i, \sum_{I=1}^{N} x_{ij} \leq M_j, x_{ij} \geq 0\right\};$$

$$F_k(x) = \sum_{i=1}^{n}\sum_{j=1}^{m}(\gamma_k d_j + b_{ijk})x_{ij}, \text{ for } k = 1, \ldots, l.$$

Then for the environmental impact *k*, for $k = 1, \ldots, l$, we can define the following problem:

Problem P_k: determine an allocation/distribution of the total amount of each waste fraction to the destinations, in order to minimize the environmental impacts *k*. That is: Determine min $\{F_k(x) : x \in X\}$.

Let $z_k^* = \min \{F_k(x) : x \in X\}$. The problem which we are interested in can be defined as follows.

Problem $P_{general}$: determine an allocation/distribution of the total amount of each waste fraction to the destinations, in order to minimize multiple environmental impacts.

A problem as $P_{general}$ is encoded in literature as a problem of "multi-objective (linear) programming" (Simeone 2002). The ideal vector of the objectives is the vector $(z_1^*, z_2^*, \ldots, z_l^*)$. If there were no conflicts among the objective functions (i.e., among the $F_k(x)$), a trivial solution to the problem would be the one obtained by solving the *l* optimisation problems one at a time (i.e., the problems $P_1, P_2, \ldots, P_l$), thus obtaining as a solution the ideal vector $(z_1^*, z_2^*, \ldots, z_l^*)$. However, generally there may be conflicts among the objective functions.

The following definition (Miettinen 1999; Simeone 2002), helps us to determine what solving a problem like this means.

Definition: A feasible solution x of P_{general} (i.e., $x \in X$) is said to be efficient or equivalently a Pareto-optimum of P_{general} if there is no feasible solution y of P_{general} (i.e., $y \in X$) such that: $F_k(y) \leq F_k(x)$ for each $k \in \{1, \ldots, l\}$, and $F_h(y) < F_h(x)$ for some $h \in \{1, \ldots, l\}$.

Then solving P_{general} means determining an efficient solution of P_{general}. Efficient solutions of P_{general} can be multiple and each of them may favour one specific goal (i.e., a certain impact category) rather than another. Therefore, the resolution of the problem takes on a subjective character, namely the decision maker has a central role in choosing the most efficient solution. Some of the methods used to find an efficient solution in multi-objective programming are described below. These methods share the same basic idea, namely to transform the original problem into one with a single objective function. Note that the methods of solution below assume that all the functions $F_k(x)$ are expressed as real numbers referred to the same unit of measurement, while the units of measurement for indicators of environmental impacts are generally different from each other. Therefore, we must proceed to the normalization of the functions $F_k(x)$, per $k = 1, \ldots, l$.

2.4 Solution Methods

Some solution methods suitable for our purpose are reviewed hereafter.

1. *Methods without preferences:* The decision maker has no preference, so the function generates any efficient solution. This method has two types of approach. The 1st type minimizes the sum of the impacts. That is:

   – Determine min $\left\{ \sum_{k=1}^{l} \left| F_k(x) - z_k^* \right| : x \in X \right\}$ *(1st type)*

Instead, the method without preferences of the 2nd type minimizes the deviation from the optimum values:

   – Determine min max $\left\{ \left| F_k(x) - z_k^* \right| (\text{with } k = 1, \ldots, l) : x \in X \right\}$ *(2nd type)*

2. *A posteriori methods:* The decision maker specifies his preferences only after all the efficient solutions have been generated. The set of all efficient solutions, called “efficient frontier,” can be calculated using (fast) algorithms (Simeone, 2002). However, the generation of all efficient solutions can be expensive, so these methods may not be feasible.
3. *A priori methods:* The decision maker specifies his preferences before an efficient solution is generated, so an efficient solution is generated based on information obtained from the decision maker. Methods of this type are:

   – Method of weights.

A relative weight (from 0 to 1) is assigned to each impact category under study according to the importance that we want to give to it, compared to the others. Then we proceed to the minimization. In general: a weight w_k is associated to each objective function $F_k(x)$, (the weights are normalized, i.e. $\sum_{k=1}^{l} w_k = 1$), and the following problem is solved: Determine min $\left\{\sum_{k=1}^{l} w_k F_k(x) : x \in X\right\}$.

– ε-constraints method.

An objective function $F_h(x)$ is selected from among the objective functions, then all the other functions $F_k(x)$ are turned into constraints (with $k = 1, \ldots, l$, with $k \neq h$), fixing upper bound ε_k on their values, and the following problem is solved:

Determine min $\left\{F_h(x) : F_k(x) \leq \varepsilon_k (\text{with} k = 1, \ldots, l, \text{ with } k \neq h), x \in X\right\}$.

– Lexicographic method

In the lexicographic method we assign an order of relative importance to the impact categories. Without loss of generality this is: $F_1(x)$, $F_2(x)$, …, $F_l(x)$.

Then we solve the following problem:

(P_1, lex) min $\left\{F_1(x) : x \in X\right\}$.

If the problem (P_1, lex) has only one solution then this is the solution of the problem P_{general}, and the algorithm ends. Otherwise, it proceeds by calculating the function with respect to the second impact category by imposing as a constraint that the solution does not worsen the value of the previous one, and so on.

That is, set x_1 = argmin of the (P_1, lex) problem, we solve the following problem:

(P_2, lex) min $\left\{F_2(x) : x \in X, F_1(x) \leq F_1(x_1)\right\}$.

If the (P_2, lex) problem has only one solution, then this is the solution of P_{general} problem, and the algorithm ends. Otherwise, it proceeds as before, by selecting the next impact category in order.

This is, at the generic step $k < l$, set x_h = argmin of the (P_h, lex) problem for $h = 1, \ldots, k\text{-}1$, we resolve the following problem:

(P_k, lex) min $\left\{F_k(x) : x \in X, F_h(x) \leq F_h(x_h) \text{ per } h = 1, \ldots, k-1\right\}$.

In practice, the solver finds an optimum solution but does not say if it is the only one. For this reason it is necessary to proceed with the very last step, which will necessarily provide a desired solution.

4. *Interactive method:* The decision maker specifies his preferences as efficient solutions are generated, guiding the resolution to the most acceptable efficient solution for him. The general scheme of this method is as follows (Miettinen 1999): (1) find an initial feasible solution; (2) present the solution to the decision maker; (3) if the solution found is good, then STOP. Otherwise, based on the information obtained, go back to point (2). This class of methods would seem to include different types of methods, since they depend on a process determining an efficient solution (there are different types of processes) (Minciardi et al. 2008).

2.5 Application

A simulation was made in order to test the validity and suitability of models.

Five waste fractions, with their relevant yearly amounts, were considered: organic (60,000 t), paper (50,000 t), glass (30,000 t), plastics (10,000 t), mixed waste (550,000 t). Six destinations, with their relevant annual capacities, were assumed: composting (25,000 t); paper recycling (14,000 t), glass recycling (20,000 t), plastics recycling (6,000 t); landfilling (170,000 t); incinerating (130,000 t). In order to assess the impact of transport, average distances between the collection centre (common to all waste fractions) and the various plants, and return, were assumed (90 km, 85 km, 80 km, 50 km, 60 km, 150 km, respectively). The transport from urban collection to the collection centre were excluded. The amount of waste generated was estimated based on average data for the Abruzzo region (Italy) and the annual capacities of plants were assumed according to the relevant national average capacities. Since all waste must be allocated, a residual destination identified as a landfill outside the region (600 km) was provided. For this plant a very high capacity was assumed that would be able to receive all the excess waste (99,999,999 t). Unit impact factors (IF) for the various waste fractions and disposal/treatment options were assessed by using the CML 2001 method with SimaPro 7.2 and Ecoinvent database. IF are not reported here, for lack of space. For the sake of this simulation, only 3 impact categories were chosen, notably: Global Warming (GWP), Acidification (AP) and Eutrophication (EP). Environmental impact indicators were normalized using the default settings. LP problems were solved by standard Excel 2010. Method without preferences, method of weights and lexicographic method were chosen to solve the MLP model.

2.6 Simulation with One Impact Category

The solution of the LP model described in Sect. 2.2 allocates the waste fractions as described in Table 2.1 in order to minimize the GWP.

In particular, as regards the organic waste we can see that no waste is allocated to the incinerator, even though, according to the relevant unit IF, this is the best destination for organic waste, when transport is not taken into account (indeed, a negative IF—i.e., an environmental "credit"—is recorded for the incineration of organic waste). The reason for the results shown in Table 2.1 is probably that the incinerator plant is located further away than the landfill and the composting plant. Therefore, the impact of transport affects the overall result to such an extent that other destinations are preferred despite the impact of the relevant processes being higher. Furthermore, in the optimal solution generated the mixed waste is allocated by preference to the incinerator, probably because the total amount of impacts for this option is the lowest. Then the model allocates this fraction to the landfill that is closer than the residual destination.

Table 2.1 Allocation of waste fractions to the destinations that minimize the GWP (t)

	Organic	Paper	Glass	Plastics	Mixed waste
Recycle	25,000	0	20,000	6,000	–
Landfill	35,000	0	10,000	4,000	121,000
Incineration	0	50,000	0	0	80,000
Residual destination	0	0	0	0	349,000

Table 2.2 Allocation of waste fractions to the destinations that minimize the AP (t)

	Organic	Paper	Glass	Plastics	Mixed waste
Recycle	0	14,000	20,000	6,000	–
Landfill	60,000	0	10,000	0	100,000
Incineration	0	36,000	0	4,000	90,000
Residual destination	0	0	0	0	360,000

Table 2.3 Allocation of waste fractions to the destinations that minimize the EP (t)

	Organic	Paper	Glass	Plastics	Mixed waste
Recycle	25,000	14,000	20,000	6,000	–
Landfill	35,000	0	10,000	0	125,000
Incineration	0	0	0	0	130,000
Residual destination	0	36,000	0	4,000	295,000

When observing allocations that minimize the AP (Table 2.2) we can already see some differences compared to the solution that minimizes GWP (Table 2.1). For AP, incineration should be preferred for the organic fraction because of the negative IF (environmental credit) if transport is not included. However, the shorter distance between the landfill and the collection centre makes landfilling environmentally better. Moreover, the model probably prefers to allocate the mixed waste to the incinerator in this case too because this is the best option for this fraction.

In the case of EP, as shown in Table 2.3, we can highlight a different distribution of waste fractions compared to the previous impact category. It should be noted that the model prefers to fill the incinerator plant up to the maximum capacity to burn the mixed waste rather than the other fractions because for this fraction this is the best allocation. Furthermore, the landfill is saturated with the mixed waste. At this point, paper and plastic must be allocated to the residual destination because the benefits from sending them to the landfill or incinerator are lower than the benefit that we can achieve by sending all mixed waste to the residual destination, or probably because the incinerator plant is fully exploited by mixed waste.

This explains why the model allocates the waste in a way that differs from what would be expected by observing the IF alone and despite the fact that there are individual solutions with minimal impact or even environmentally beneficial impact.

Table 2.4 Optimum allocation of waste fractions to the various destinations according to the method of weights (weights: GWP=0.5, AP=0.3, EP=0.2) (t)

	Organic	Paper	Glass	Plastics	Mixed waste
Recycle	25,000	14,000	20,000	6,000	–
Landfill	35,000	0	10,000	0	125,000
Incineration	0	36,000	0	0	94,000
Residual destination	0	0	0	4,000	331,000

Table 2.5 Optimum allocation of waste fractions to the destinations according to the method without preferences of 1st type (t)

	Organic	Paper	Glass	Plastics	Mixed waste
Recycle	25,000	14,000	20,000	6,000	–
Landfill	35,000	36,000	10,000	0	89,000
Incineration	0	0	0	0	130,000
Residual destination	0	0	0	4,000	331,000

2.7 Simulation with MLP Model

Three methods of solution were used: method of weights, method without preferences and the lexicographic method.

For the resolution of the MLP model with method of weights, the highest weight was assigned to GWP (0.5), while weights of 0.3 and 0.2 were assigned to AP and EP respectively. Leaving aside the other allocations, it should be noted that the solution prefers to allocate the plastics fraction to the residual destination rather than to the landfill or incinerator plant, despite the greater distance of the former, because it is more convenient from an environmental point of view to fill the incinerator and landfill with the mixed waste. Hence, the only available option for the plastics fraction is the residual destination (Table 2.4). When very different weights are assigned (i.e., 0.1 to GWP, 0.3 to AP and 0.6 to EP), the results as regards allocation do not change greatly. In fact, in this case, sending the plastics fraction to the landfill is preferred to sending the mixed waste to the landfill.

With the method without preferences of the 1st type we obtain the distribution shown in Table 2.5. Recycling plants are fully exploited by their relevant recyclable fractions. The surplus is diverted to the landfill, with the exception of the plastics fraction, which is allocated to the residual destination to make way for mixed waste at the incinerator and the remaining portion of the landfill. This is probably because the IF related to the EP of the plastics fraction sent to the incinerator is greater than the IF of mixed waste fraction to the same process. Thus, the plastics fraction is sent to the residual destination as the fact that the other plants have reached full capacity means there is no alternative. In the case of the method without preferences of the 2nd type, the function gives us the same allocation as the 1st type. The only difference is that the surplus differentiated fraction in the first case is sent to the landfill, while in the second case it is sent to the residual target. In this case too there is a domino effect.

Table 2.6 Optimum allocation of waste fractions to the destinations with the lexicographic method in the first hypothesis (t)

	Organic	Paper	Glass	Plastics	Mixed waste
Recycle	25,000	0	20,000	6,000	0
Landfill	0	0	10,000	0	160,000
Incineration	0	50,000	0	0	80,000
Residual destination	35,000	0	0	4,000	310,000

According to the description introduced in Sect. 2.4 for the lexicographic method we at first hypothesized the following order: GWP (constraint), AP (1st step), EP (2nd step).

The optimum solution of the function calculated for GWP is set as a constraint to calculate the optimum in the 1st step for AP. In the 2nd step we calculated the optimum for EP by setting the optimum of AP as a constraint. In fact, by observing the results of the 2nd step, that in this case coincides with the final results (Table 2.6), we note that these differ little from the solution obtained in the 1st step. This can be explained by the domino effect. In this case we obtained a very similar distribution to that obtained by calculating the optimum for the GWP (see Table 2.1), even though the organic fraction and the plastics that cannot be recycled are allocated to the residual destination rather than the landfill to make way for the mixed waste fraction. This is because, by imposing that GWP should not get worse as a constraint, results in the interaction with the other impact categories changed the final solution only slightly.

In another simulation the order followed was: AP (constraint), EP (1st step), GWP (2nd step). Once again the difference with the solution obtained from the LP model for AP (Table 1.2) is minimal: the organic fraction is diverted into the residual destination instead of being allocated to the landfill. Obviously this leads an offsetting of the values of mixed waste sent to the various destinations.

2.8 Conclusions

When IWM scenarios are to be selected, LCA can help decision makers identify the best one from an environmental point of view. A limitation of implementing LCA alone is that a limited number of alternative scenarios—extracted from a potentially infinite set of available scenarios—are usually compared. In this way, scenarios better than those ultimately selected for comparison might be neglected *a priori*. An LP model and MLP model were proposed to be integrated with LCA in order to identify the most environmentally-friendly scenario(s) from a potentially infinite set of scenarios, given the limitations of the context under study. The proposed models have also been applied to a hypothetical and plausible context to test their validity and suitability. The results obtained show how the results may vary greatly according to the method adopted. In addition, some factors such as transport, which is conditioned by the distance of the treatment/disposal plants from the waste collection centre, can significantly affect the system management. The simulation was performed by considering a steady state, however the models proposed can also be used to assess how the optimum solution may change as a result of new investments in additional plants. A further step of application to a real context is expected. An IWMS has been identified and data will be collected. This phase of the project allows us to understand whether the model developed can be applied to a more complex system. Furthermore, it will allow us to understand whether it may be adapted to meet different requirements as they arise over time.

References

Fukushima Y, Hirao M (2002) A structured framework and language for scenario-based life cycle assessment. Int J LCA 7:317–329. doi:10.1007/BF02978679

Mercuri L, Raggi A (2004) A proposal for a mathematical model to identify optimal scenarios for waste management systems. Proceeding. Sixth International Conference on EcoBalance, Tsukuba, Japan, 25–27 October, pp. 481–482

Miettinen K (1999) Nonlinear multi-objective optimization. Kluwer Academic Publishers, Boston

Minciardi R, Paolucci M, Robba M, Sacile R (2008) Multi-objective optimization of solid waste flows: environmentally sustainable strategies for municipalities. Waste Manag 28:2202–2212. doi:10.1016/j.wasman.2007.10.003

Simeone B (2002) Preferenze, decisioni multicriterio e scelte collettive: tre risultati chiave, in modelli e algoritmi per l'ottimizzazione di sistemi complessi, Atti della Scuola CIRO, Pitagora, Bologna, Italy

Tascione V, Raggi A (2010) Rassegna critica degli approcci relativi all'identificazione degli scenari nell'applicazione della Life Cycle Assessment alla gestione integrata dei rifiuti: risultati preliminari. In: Morselli L (ed) Ambiente-Economia. Nel cuore delle azioni. Proceedings 'Ecomondo 2010', Rimini, 3-6 November, Maggioli, Santarcangelo di Romagna, pp. 1209–1214

Tascione V, Raggi A (2012) Identification and selection of alternative scenarios in LCA studies of integrated waste management systems: a review of main issues and perspectives. Sustainability 4:2430–2442

Chapter 3
Urban Metabolism: Many Open Questions for Future Answers

Giuseppe Ioppolo, Reinout Heijungs, Stefano Cucurachi, Roberta Salomone and René Kleijn

Abstract Currently, the Urban Metabolism (UM) approach, as deducted from international literature, has been applied several times to assess and describe urban flows and impacts related to them, using different tools such as Material Flow Account (MFA). Although very useful, UM is not a systematic approach in urban development, and, for this reason, the authors of this chapter introduce MFA into the urban planning model, designing a research project in the field of Industrial Ecology (IE) applied in an urban context. This chapter presents the preliminary considerations of the first conceptualization phase of this project.

Keywords Urban metabolism · Material flow account · Industrial ecology · Strategic environmental assessment · Eco-governance

3.1 Introduction

Artificial land cover increased by 3.4 % in Europe between the years 2000 and 2006 (EEA 2009): this is by far the largest proportional increase in all land use categories. Although artificial cover accounts for just 4 % of the EU land area, the fact that it is dispersed means that more than a quarter of the EU territory is directly affected by urban land use (the residential, commercial and industrial land use of cities, so also called urban areas) (EEA 2009). Between 2007 and 2025, the urban areas of the world are expected to gain 1.3 billion people, including 261 million in China and 197 million in India, which account together for 35 % of the total increase (DESA 2008). By 2020, approximately 80 % of the Europeans will be living in urban areas. Furthermore, this urban growing is neither uniform nor constant, and often occurs

G. Ioppolo (✉) · R. Salomone
Department SEAM, University of Messina, Piazza S. Pugliatti 1, 98122, Messina, Italy
e-mail: giuseppe.ioppolo@unime.it

R. Heijungs
Department of Econometrics and Operations Research,
VU University Amsterdam, Amsterdam, The Netherlands

R. Heijungs · S. Cucurachi · R. Kleijn
Institute of Environmental Sciences, University of Leiden, Leiden, The Netherlands

R. Salomone, G. Saija (eds.), *Pathways to Environmental Sustainability*,
DOI 10.1007/978-3-319-03826-1_3,

in a scattered way throughout the European countryside, determining a phenomenon called urban sprawl (EEA 2009; ISPRA 2011; Hasse and Lathrop 2003). Obviously, the extension of urban areas offers benefits, allowing people to enjoy more living space, single-family houses and gardens; but it can also create negative environmental, social and economic impacts for European cities and countryside, in particular in the case of low density and scattered urban sprawl. The urban sprawl is not only a problem for the absence of a long term sustainable policy for the city development, but above all, the impact drive force that produces increasing energy demand, human health problems, declining stocks of natural resources, and, from a social perspective, exacerbated social and economic divisions (Catalàn et al. 2008).

Historically, the fields of urban planning and city management have been concerned with zoning, land use, growth channeling and with the systematic provision of infrastructure (Gappert 1989). However, in the next decades, urban planning and city management will have to include the impact of cities on the environment and their contribution to global environmental solutions, as traditional urban management practices lack the comprehensiveness required by sustainability (Jones et al. 2002). In the perspective of these outlooks, this paper uses a different point of view for the analysis of territory, focusing on an urban system. The urban system is a territorial ecological system strongly affected by human activities. Its observation takes into account the kind of exchanges and fluxes which were generated within the system and between the system and its environment. A possible approach to study the complex exchanges of the urban system is based on the analogy between city system and biological organisms and ecosystems (Decker et al. 2000). Applying these assumptions, social scientists extended the concept of metabolism from biological science to the analysis of human systems. According to Chun-Lin et al. (2009), socio-economic metabolism, also known as Urban Metabolism (UM), has been extensively applied to explore resource consumption, asset accumulation, waste emissions, and complex processes of land use change in a socio-economic system, (Wolman 1965; Rappaport 1971; Fischer-Kowalski 1998).

UM can be defined as the extraction of resources from the socio-economic system, the consumption of resources from the outer system, the accumulation of assets in the system and the emission of waste by the system (EUROSTAT 2001; Brunner and Rechberger 2004). The term UM transposes to the urban scale the concept of Industrial Metabolism, based on the notion that economic systems can be analyzed in terms of material and energy transformations with metabolic pathways that evolve over time (Ayres and Simonis 1994; Fischer-Kowalski 1998; Fischer-Kowalski and Haberl 2007; Siciliano 2012). Under this approach, a territory may be seen as an "open system that exchanges with other territories (goods, materials, energy, etc.), and with environmental compartments (air, soil, water) as well as with flora and fauna" (Calame 2009), and "cities, as an organism, transform raw materials, fuel, and water into the built environment, human biomass and waste" (Decker et al. 2000).

In order to study the UM, several methods and tools are introduced to assess and manage the urban growth and transformation (e.g. according to Hodson (2012) the Material Flow Analysis (MFA) applied at an urban scale). Indeed, MFA investigates resource-oriented problems (Loiseau et al. 2012) exploring the physical relations between the actions of individuals and the relative environmental impacts. However, the complexity of the urban context need an interdisciplinary approach. For instance, Broto et al. (2011), use the UM concept to define a functional analogy (i.e. city as a living organism under a physiological perspective), to adopt a form analogy (i.e. city as a system of social and ecological components in a morphological perspective), and to foster a common pathway that optimizes the urban flows from both an environmental and a political and economic perspectives.

In order to combine the aforementioned tools, we identified the need of providing public decision makers and citizens with a comprehensive framework which would require an in-depth research articulated in the following phases:

- definition of the key factors and the key questions to introduce Material Flow Account (MFA) in the urban planning method through the UM approach;
- design of a representative model, supported by a specific literature review crossing different criteria, with the goal of integrating the UM approach and the MFA method with traditional urban planning tools (e.g. SEA, EIA) (McCluskey and João 2011), using data integration (i.e. calibration), scenario creation and comparison, and a set of aggregate indicators (Chaker et al. 2006);
- model testing on a pilot urban system (after setting the model on the specific city dimension);
- definition of possible sustainable middle-long term actions and scenarios to plan the city system.

According to Reijnders (1998), the main problem to deal with implementing the research in a urban context is to answer a core question: *what we have to reduce to achieve a sustainable resource throughput?* This chapter carries out the preliminary considerations on this main problems and highlights the strategic key points that should be firstly tackled.

In the following sections, the proposed contribution will be described. In the method and tool section, the authors aim to define the common criteria guiding the approaches of Industrial Ecology (IE) to the urban development, defining question points and key tools to support the model (next research phase out of this article). The discussion section summarizes possible strengths and weaknesses to take the project forward. Finally, in the conclusion section several critical aspects and open research questions are highlighted.

3.2 Method and Tool: Industrial Ecology for Urban Sustainability

Labor, capital and technology interact with each other in an industrial ecology system, as they do in an urban system. This analogy eases the move from the concept of Industrial Metabolism to the concept of UM. In order to achieve a better understanding of the structure of an anthropogenic system, such as urban areas, and in order to define new cycles into the urban system, Tibbs (1993) and Ayres and Ayres

(2002) use the term "industrial ecosystem" to underline the analogies of its characteristics to the one of a natural ecosystem. To complete this concept it is necessary to introduce a time-scale, that in a techno-sphere system is relevant and not unlimited, and which allows the analysis of the interactions happening among all actors (competition and/or co-operation types).

The IE principles highlighted in Tibbs (1993) and Ayres and Ayres (2002), could be considered as main support to make environmental considerations as a part of every strategy, design, production and product end-of-life activity, through the re-engineering of processes and of activities in an industrial eco-system. Studies in the field of IE have clearly shown that side-effects of environmental policy measures are often caused by changes in the closely interconnected material flows in the society (De Marco et al. 2001; Kleijn 2012). According to Kleijn (2012), the scarcity of resources, the growing material and energy consumption requirements, the dematerialization and substitution approach for sustainable development, are the clear outlines for a sustainable society. They can be phrased in terms of three major transitions:

- a transition from fossil fuels to renewable energy sources;
- a transition from linear material flows (from raw materials to waste) to a closed-loop materials economy;
- a transition from the exploitation of nature and biodiversity to its protection.

More efficient and dematerialized processes (Hinterberger et al. 2003), with a reduction of input (e.g. due to a better intra-boundaries use of flows, etc.), are necessary to achieve a more sustainable new urban system. A set of general objectives of this vision are:

- maximizing the efficiency of management by using a decision support system integrated in a Strategic Environmental Assessment (SEA) in order to follow the European Directive 2001/42/EC (EU 2001, Loiseau et al. 2012);
- analyzing urban systems in relation to the surrounding territorial context, by using a positive integration of tools and avoiding overlapping in the analysis phase in order to prevent contrasting results;
- seeking new organizational structures for urban systems, by integrating the improvement of flow management from the outset (MFA supporting planning and program phases, eventually, combined with life cycle tools);
- building a common vision to apply a shared strategy linked to urban sustainability and quality of life, through the valorization of the behaviors of all components of the context (e.g. which actors exchange intra the/with others system boundaries).

Following the criteria of the field of IE, few entire urban projects were conducted, such as the eco-city of Guiyang, China (Shi et al. 2003) or in Japan (van Berkel et al. 2009). The level of complexity of a urban system is very high. Therefore, to investigate and answer to the above mentioned main question, it is firstly necessary to define a few common baseline criteria, as summarized in the following points:

- defining the specific matters of urban resource management for the context analyzed (e.g., definition of boundaries of model at the scale of town or district/neighborhood); definition of significant flows as energy, water, building material (e.g. for the construction, use and disposal phases) related to inhabitants behaviors; definition of auto-production and recovery (e.g. local plants of energy production such as solar, mini/micro wind energy-system, biogas production, etc., or the house water cycle system);
- identifying and applying the methodologies and indicators to measure a circular UM. These indicators should reflect social implications (e.g., urban sustainability, for which "ad-hoc" set indicators were used, defining an environmental label in order to permit a benchmarking intra system—from buildings, or from neighborhoods—and inter systems—from cities—Scipioni et al. 2009; van Berkel et al. 2009);
- defining a flexible model able to link perspectives of improving urban resource efficiency and environmental quality scenarios with the targets of urban governance, to integrate environmental planning in a urban context. From a common knowledge based on shared information, it is, in fact, possible to improve the behaviors in a more responsible way, building a new eco-governance with common vision, strategy and objectives/targets (Ioppolo et al. 2012).

The definition of quantitative dematerialization and recycling targets, at a local (urban) scale, possibly using MFA is a fundamental step of the analysis. Kovanda et al. (2009) and Kennedy et al. (2011) defined the state of art summarizing the main UM studies available in literature. Several empirical MFA studies on regional or local levels have been carried out, but they remain few compared to the large number of MFA studies on the national level.

3.3 MFA in a Urban Metabolism Approach

The MFA methodology is summarized in the following. One of the main common methods to apply MFA was developed by the Statistical Office of European Communities (EUROSTAT 2001) and extensively used at different territory scales (e.g., Vienna and Swiss Lowlands by Hendrics et al. (2000); Singapore by Schulz (2007); Paris and its suburbs by Barles (2009); or with a generic point of view as Deilmann (2009).

The application of the method provides three main areas to conduct an analysis (van der Voet 2002): definition of the system, quantification of the overview of stocks and flows and the interpretation of results. These three areas incorporate the six following steps (Brunner and Rechberger 2004):

- Definition of the system—definition of goals and selection of monitoring indicators; system definition in closed boundaries, including scope and time frame;

- Quantification of the overview of stocks and flows—identification of relevant flows, process and stocks; design of all flows in a comprehensive scheme; mass balancing;
- Interpretation of results—interpretation of result and summary, evaluating the robustness of the overview quantification and translating this into policy-relevant terms (van der Voet et al. 1999).

The indicators used are standard indicators, developed in the framework of material and energy flow analysis on the level of national economies (EUROSTAT 2007; Haberl et al. 2004), and they express the amounts actually used by a social system during the course of a year (metabolic rate), while the stocks represent the system size. The main indicators built up from MFA are (Brunner and Rechberger 2004):

- Direct Material Input (DMI): it is the sum of domestic extractions and imports;
- Direct Material Consumption (DMC): it adds domestic extractions and imports and subtracts exports. It symbolizes the domestic waste potential of a system;
- Physical Trade Balance (PTB): it is equal to importations less exportations. It is the opposite of the trade balance;
- Net Addition to Stock (NAS): it measures the physical growth of an economy, i.e., the materials that are added to stocks every year;
- Domestic Processed Output (DPO): it measures all the materials which are released into the environment (air and water emissions, industrial and domestic wastes which are sent to waste disposal sites as well as diffuse flows);
- Indicators that are based on ratios have also been built (EUROSTAT 2009):
- Area Intensity: is the ratio between domestic extractions (DE or DMC) and total land area. It provides information on the relationship between the scale of the physical economy and its local environment;
- DE/DMC: it indicates the degree of dependency of the physical economy for raw materials supply on the study area.

During the model definition phase, in each applicative research, contradictions emerge as (Hammer et al. 2003):

- direct flows (actual weight of products without a chain dimension view) versus indirect flow (wide perspective including all up-stream resource requirements);
- used versus unused (all used materials, exchanged within the economic system, are taken into account as the amount of extracted resources, leaving out of the economic system, as unused extraction, the physical market externalities);
- domestic versus rest of the world (ROW), represent another debated problem due to the origin and/or destination of the flows, and to the consequence geography assignment of the related environmental burdens.

In previous studies on UM there is a problem of comparability, due to the subjectivity of the approaches undertaken, as most of all lack of basic tools to translate their results into urban policy suggestions. From a biological point of view, the city consists of a mosaic-like multiplicity of biotopes. As a rule, they are clearly demarcated and internally relatively homogeneous. Urban biotope mapping provides a good overview of the biotope types and vegetation structures in a city. Probably, this remains one way for the future research, and the definition of urban typologies-biotopos-urbantopos (e.g., the location of the districts/neighborhoods—central, transition or peripheral) becomes the determining base for the establishment of thresholds in the identification of critical impacts of energy and matter flows and also to design strategic actions. The urban typologies will be used later to tackle the selection of indicators, and their corresponding thresholds, to identify critical impacts on matter, energy, water, food and CO_2 flows.

3.4 Discussion on Future Perspective

The city material consumption was already discussed, but often analyzed as a specific component rather than a comprehensive UM, without a cross-sectorial approach to define a comparable quality of an urban system. Furthermore, the intra flow relations are complex and need a dynamical approach. For instance, Hu et al. (2010), using the stock dynamics models proposed by Müller et al. (2004) and Müller (2006), investigated the long-term metabolism (using parameters as population, lifestyle, technology, and product lifetime) for the floor area and selected a construction material (i.e. steel) in China's residential building stock (rural and urban housing). From this experience emerges that in a country context there is a significant amount of relevant data available from statistical offices and from material input and trade. On a regional and local level, data availability in general is much smaller and the gathering step could be time-consuming and not relevant. The methodology presents both strengths and weaknesses. MFA is a consistent method, well established and transparent, flexible and proportionally comprehensible, with a wide and long history of international applications in different fields. It represents a base to develop a common international comparison system, using the indicators compiled by Eurostat and the OECD. UM could contribute as the common theoretic base in which to define possible scenarios of economy-wide material flow account by local (urban) to international/global contexts. In a perspective of policy integration, MFA, does not distinguish among different qualities of flows, which are assessed in tonnes as the lowest common denominator. An eco-efficiency ratio is not introduced, as well as an urban inventory to take in account the main impact categories (Seppälä et al. 2005), ignoring several resource uses such as water consumption and land use. Above all, remain as problem, the absence of an integrated dataset per city context due to the scarce availability of information about UM and MFA in local areas.

3.5 Conclusions

The state of the art, as the scientific literature shows, seem to be promising in overcoming the above mentioned key weaknesses and seem to achieve the common assumptions earlier on described. According to Baccini and Oswald (2008) we could adopt the four principles for the redesigning and the reconstruction cities (shapability, sustainability, reconstruction, and responsibility), and assess, as major components of UM, water, food (biomass), construction materials, and energy. An idea, that reflects this principles, and tackles the main weaknesses (see previous Sect. 3.3), is to create a database for each city as e.g. defined in Directive 2002/91/EC for managing the energy building consumption (Andaloro et al. 2010), applying a Geo Information System as in a cadastral (the information on owner, size, map, position, etc., associated with the related average building consumption (Reiter and Marique 2012). Nevertheless, putting in actions these key factors actually remain an open question, and to define tangible solutions, great financial supports and a great strict collaboration between, scientists, public system and private system are needed. In this way municipality or local authority could represent a guide to achieve this innovation. The challenge ahead is to design a new model for city, using the UM approach and supporting a specific ad-hoc city label with environment and energy each other combined.

References

Andaloro APF, Salomone R, Ioppolo G, Andaloro L (2010) Energy certification of buildings: a comparative analysis of progress towards implementation in European countries. Energy Policy 38:5840–5866

Ayres RU, Simonis UE (1994) Industrial metabolism: restructuring for sustainable development. United Nations University Press, Tokyo:376

Ayres RU, Ayres LW (2002) Handbook of industrial ecology. Edward Elgar, Cheltenham

Baccini G, Oswald PF (2008) Designing the urban: linking physiology and morphology. In: Hadorn H (ed) Handbook of transdisciplinary research. Springer, Dordrecht

Barles S (2009) Urban metabolism of Paris and its region. J Ind Ecol 13:898–913

Broto VC, Allen A, Eriksson A (2011) Urban metabolism at UCL. A working paper. UCL Environmental Institute, London. http://www.ucl.ac.uk/environment-institute/research/metabolism. Accessed 10 Oct 2012

Brunner PH, Rechberger H (2004) Practical handbook of material flow analysis. Lewis, Boca Raton

Calame P (2009) Le territoire, acteur pivot du XXIe siècle. In: Calame P Essai sur l'Oeconomie. Editions Charles Léopold Mayer, Paris. http://www.i-r-e.org/IMG/pdf/partie_II_chap_5.pdf. Accessed 10 Oct 2012

Catalàn B, Sauri D, Serra P (2008) Urban sprawl in the Mediterranean? Patterns of growth and change in the Barcelona metropolitan region 1993–2000. Landsc Urban Plan 85:174–184

Chaker A, El-Fadl K, Chamas L, Hatjian B (2006) A review of strategic environmental assessment in 12 selected countries. Environ Impact Assess Rev 26:15–56

Chun-Lin L, Shu-Li H, Shih-Liang C (2009) Synthesis and spatial dynamics of socio-economic metabolism and land use change of Taipei metropolitan region. Ecol Model 220:2940–2959

De Marco O, Lagioia G, Pizzoli Mazzacane E (2001) Materials flow analysis of the Italian economy. Ind Ecol 4:55–70

Decker H, Elliott S, Smith FA, Blake DR, Sherwood Rowland F (2000) Energy and material flow through the urban ecosystem. Annu Rev Energy Environ 25:685–740

Deilmann C (2009) Urban metabolism and the surface of the city. In: German annual of spatial research and policy. Guiding principles for spatial development in Germany. Springer, Berlin Heidelberg

DESA (Department of Economic and Social Affairs) (2008) World urbanization prospects. The 2007 revision. United Nations, New York
EEA (European Environment Agency) (2009) Ensuring quality of life in Europe's cities and towns. Office for Official Publications of the European Communities, Luxembourg
EU (European Union) (2001) Directive 2001/42/EC of the European Parliament and of the Council of 27 June 2001 on the Assessment of the Effects of Certain Plans and Programmes on the Environment
EUROSTAT (2001) Economy-wide material flow accounts and derived indicators. A methodological guide. Statistical Office of the European Union, Luxembourg
EUROSTAT (2007) Economy-wide material flow accounting. A compilation guide. European Statistical Office, Luxembourg
EUROSTAT (2009) Economy wide material flow accounts: Compilation guidelines for reporting to the 2009 Eurostat questionnaire. European Statistical Office, Luxembourg
Fischer-Kowalski M (1998) Society's metabolism. The intellectual history of material flow analysis, part I: 1860–1970. J Ind Ecol 2:61–78
Fischer-Kowalski M, Haberl H (2007) Socio-ecological transitions and global change: trajectories of social metabolism and land use. Edward Elgar, Cheltenham, p 263
Gappert G (1989) A management perspective on cities in a changing global environment. In: Knight RV, Gappert G (eds) Cities in a global society. Urban Affairs Annual Reviews, Beverly Hills. pp 312–325
Haberl H, Fischer-Kowalski M, Krausmann F, Weisz H, Winiwarter V (2004) Progress Towards Sustainability? What the conceptual framework of material and energy flow accounting (MEFA) can offer. Land Use Policy 21:199–213
Hammer M, Giljum S, Barbigli S, Hinterberger F (2003) Material Flow Analysis on the Regional Level: Questions, Problems, Solutions. NEDS Working Paper. No. 2, Hamburg
Hasse JE, Lathrop RG (2003) Land resource impact indicators of urban sprawl. Appl Geogr 23:159–175
Hinterberger F, Giljum S, Hammer M (2003) Material flow accounting and analysis (MFA) A valuable tool for analyses of society-nature interrelationships. Sustainable Europe Research Institute (SERI), Wien. http://seri.at/wp-content/uploads/2009/09/Material-Flow-Accounting-and-Analysis-MFA.-Encyclopaedia-of-the-International-Society-for-Ecological-Economics-ISEE.pdf. Accessed 10 Oct 2012
Hu M, Pauliuk S, Wang T, Huppes G, van der Voet E, Müller DB (2010) Iron and steel in Chinese residential buildings: a dynamic analysis. Resour Conserv Recycl 54:591–600
Ioppolo G, Saija G, Salomone R (2012) Developing a territory balanced scorecard approach to manage projects for local development: two case studies. Land Use Policy 29:629–640
ISPRA (2011) Qualitá dell'ambiente urbano VII rapporto annuale. ISPRA, Roma. http://www.minambiente.it/export/sites/default/archivio/biblioteca/qualita_ambiente_urbano_VII_rapporto_2010.pdf. Accessed 10 Oct 2012
Kennedy C, Pincetl S, Bunje P (2011) The study of urban metabolism and its applications to urban planning and design. Environ Pollut 159:1965–1973
Kleijn R (2012) Materials and energy: a story of linkages. Muller Visual, Leiden
Kovanda J, Weinzettel J, Hák T (2009) Domestic extraction used (DE) and physical trade in Prague and the Czech Republic: Comparison and argumentation for the best DE ratio. In: Havránek M (ed) ConAccount 2008. Urban metabolism: measuring the ecological city. Book of proceedings. Charles University Environment Center, Prague
Jones P, Tauléelo T, Kohlhase J (2002) Growing pacific town and cities: Samoa's new planning and urban management system. Aust Plan 39:186–193
Loiseau E, Junqua G, Poux P, Bellon-Maurel V (2012) Environmental assessment of a territory: an overview of existing tools and methods. J Environ Manage 112:213–225
McCluskey D, João E (2011) The promotion of environmental enhancement in strategic environmental assessment. Environ Impact Assess Rev 31:344–351
Müller DB (2006) Stock dynamics for forecasting material flows—case study for housing in the Netherlands. Ecol Econ 59:142–156

Müller DB, Bader HP, Baccini P (2004) Long-term coordination of timber production and consumption using a dynamic material and energy flow analysis. J Ind Ecol 8:65–88
Rappaport R (1971) The flow of energy in an agricultural society. Sci Am 225:117–132
Reijnders L (1998) The factor X debate: Setting targets for eco-efficiency. J Ind Ecol 2:13–22
Reiter S, Marique A-F (2012) Toward low energy cities. J Ind Ecol doi: 10.1111/j.1530-9290.2012.00533.x
Schulz NB (2007) The direct material inputs into Singapore's development. J Ind Ecol 11:117–131
Scipioni A, Mazzi A, Mason M, Manzardo A (2009) The dashboard of sustainability to measure the local urban sustainable development: The case study of Padua Municipality. Ecol Indic 9:364–380
Seppälä J, Melanen M, Mäenpää I, Koskela S, Tenhunen J, Hiltunen M-R (2005) How can the eco-efficiency of a region be measured and monitored? J Ind Ecol 9:117–130
Shi H, Moriguichi Y, Yang J (2003) Industrial ecology in China, part 1: research. J Ind Ecol 6:7–11
Siciliano G (2012) Urbanization strategies, rural development and land use changes in China: a multiple-level integrated assessment. Land Use Policy 29:165–178
Tibbs H (1993) Industrial ecology. An environmental agenda for industry. Global business network. http://www.hardintibbs.com/wp-content/uploads/2009/05/tibbs_indecology.pdf. Accessed 10 Oct 2012
van Berkel R, Fujita T, Hashimoto S, Geng Y (2009) Industrial and urban symbiosis in Japan: analysis of the Eco-Town program 1997–2006. J Environ Manage 90:1544–1556
van der Voet E (2002) Substance flow analysis methodology. In: Ayres RU, Ayres LW (eds) A handbook of industrial ecology. Edward Elgar, Cheltenham
van der Voet E, van Oers L, Guinée JB, Udo de Haes HA (1999) Using SFA indicators to support environmental policy. Environ Sci Pollut Res 6:49–58
Wolman A (1965) The metabolism of cities. Sci Am 213:179–190

Chapter 4
Water Footprint to Support Environmental Management: An Overview

Anna Mazzi, Alessandro Manzardo and Antonio Scipioni

Abstract The issue of water and its management is central to international debate: water is a scarce resource, although renewable; climate changes, agricultural, industrial and civilians use have limited the availability of this resource for future generations. In recent years, companies have shown a growing interest in consequences of water use and consumption, especially the agri-food sector companies. In this chapter, through an overview of methodological and operational approaches used in scientific literature to assess water use and consumption, we summarize their main applications in agri-food sector and we indicate future developments in the field.

Keywords Water footprint · Agri-food sector · accounting and impact assessment · ISO 14046 · Literature review

4.1 Introduction

Fresh water is a vital asset to ensure the life and the economic and social development of humanity. In recent years the issue of water and its management has become central to international debate. This interest stems from that water is a scarce resource, although renewable. Climate change and water use for agricultural, industrial and civilians have limited the availability of fresh water, threatening the accessibility of this resource for the future (IPCC 2008; UNESCO 2009; WWAP 2009).

European Union has recognized the importance of proper management of this resource by adopting new water policy aimed at water saving, protection of quality and ecosystems of watersheds, and resolution of inefficiencies of transmission networks (EU 2012).

In recent years, companies have shown a growing interest in issues related to water resources, because of the consequences that the issue has on the environmental but also economic performance of an organization (WBCSD 2006, 2009).

A. Scipioni (✉) · A. Mazzi · A. Manzardo
Department of Industrial Engineering, Quality and Environment Research Center, University of Padova, 35121, Padova, Italy
e-mail: scipioni@unipd.it

R. Salomone, G. Saija (eds.), *Pathways to Environmental Sustainability*,
DOI 10.1007/978-3-319-03826-1_4,

Particularly, food companies were the first to demonstrate a keen interest in the establishment of standardized methodologies to better understand and manage their water use and consumption. The reasons are manifold. Firstly, in the agri-food sector water consumption is higher than other sectors (CAWMA 2007; WWAP 2009). Secondly, the major multinational companies operating in agri-food industry have facilities in countries with serious problems of water scarcity (WBCSD 2009). Last but not least, some laws enacted by individual countries and the European Union stressed the need for a more conscious use of water resources (PF 2010; EU 2012).

Latest research on methodologies and tools to support better water use and management are taking place within the framework of Life Cycle Assessment (LCA). Such tools and methodologies are combined to develop new integrated models (see also Kounina et al. 2013; ISO 2012).

The goals of this chapter are: to give an overview of the methodological and operational tools developed in recent years for water use and consumption assessment, summarize the main applications in the agri-food sector and indicate future developments in the field including standardization processes at international level.

4.2 Overview of Approaches for Assessing Water Use

4.2.1 The Virtual Water Concept

The need to formulate methods to manage the water resource is derived from the main problem of water scarcity, which in turn derives from anthropogenic changes to local water availability (Pfister et al. 2009). One of the first initiatives in this area is proposed by Allan (1998), which, with the aim of transferring the availability of water resources across the continent, introduces the concept of virtual water: it measures the amount of water that can be transferred in countries with water shortages through trade of products containing water, mainly agricultural and food products.

4.2.2 The Water Footprint Accounting Concept

Building on the concept of virtual water, Hoekstra and Chapagain (2008) introduced the concept of water footprint (WF) as a method to quantify the appropriation by man of globally available fresh water. WF at product level represents the sum of all the water used in a supply chain, comprising blue, green and grey water (Hoekstra 2010; Hoekstra et al. 2011).

Blue water is defined as the volume of freshwater abstracted from rivers, lakes and aquifers; this component is preponderant in some types of cultivation, in farms and in some industrial processes. The amount of rainwater (stored in the soil as soil moisture) used by plants is referred to as green water; this component is significant in agricultural production processes. Finally, grey water accounts for the impact of pollution on water resources and represents the volume of freshwater needed to dilute pollution so that the quality of the water remains above water quality standards set by regulations; this component is especially important in industrial processes (Table 4.1).

4.2.3 The Life Cycle Impacts on Water Concept

An important evolution of the water impact assessment comes from the contribution of several authors that introduce corrective elements to the life cycle assessment (LCA) methodology for an in depth study on water resources related impacts.

LCA methodology, established in the early sixties in order to study the energetic burdens associated with certain industrial products (see also SETAC 1991; Hunt and Franklin 1996; EPA 2006), has evolved over the years to be today the most comprehensive method of environmental impacts assessment of a product, service, process or system with a life cycle perspective (ISO 2006, 2012).

The introduction of water footprint concept within the LCA methodology is intended to complement and enhance life cycle impact assessment (LCIA), and to obtain a more complete estimate of life cycle impacts on water.

Integration of methodologies to assess impacts on water is a challenging issue within the LCA framework. Several methodologies have been published to do so.

Eco-scarcity method is one of the methodological framework presented to integrate such impacts in LCIA. This method provides eco-factors for various environmental impacts including water use (Frischknecht et al. 2009). Here, water use is defined as the total input of freshwater abstracted for production or consumption. Water use is grouped into six water-scarcity categories from low (using less than 10 % of the available freshwater resources) to extreme (using more than 100 % of the available freshwater resources). For assessing the impacts of water scarcity, each category is then assigned an individual ecofactor based on the average water withdrawal-to-availability values. Furthermore, this concept considers the cause–effect chains that link freshwater type and use to potential impacts at the mid- and endpoint level and to the related area of protection of human health, ecosystem quality, and resources (Jolliet et al. 2004).

Another approach suggested by Milà i Canals et al. (2009) considers water use at the level of a river basin and includes both the source of water and type of use of freshwater in life cycle inventory. With respect to the classification seen below (blue water, green water and grey water), Milà i Canals et al. introduces a further classification: the blue water is differentiated into flow (river/lake), fund (aquifer) and stock (fossil); the water use is split into evaporative and non-evaporative use, the latter is defined as water returned to the freshwater source after its use and available for further use (Table 4.1).

The approach suggested by Pfister et al. (2009) considers only blue water and differentiates three categories of water use: in-stream water use, water consumption (where the water is no longer available in the watershed) and water-quality degradation (where the water is still available after use but with diminished quality) (Table 4.1).

Table 4.1 Terminology used in different water use assessment concepts. (Adapted from Jeswani and Azapagic 2011)

Freshwater use concept	Freshwater component	Term explanation	Literature reference
Virtual water concept	Virtual water	Amount of water content of agri-food products	Allan 1998; Pfister et al. 2009
Water footprint accounting concept	Blue water	Freshwater available in surface water bodies (rivers, lakes) and aquifers for abstraction	Hoekstra and Chapagain 2008; Hoekstra 2010; Hoekstra et al. 2011
	Green water	Rainwater (stored in the soil as soil moisture) used by plants and vegetation	
	Grey water	The volume of freshwater required to dilute pollutants so that the quality of water remains above water quality standards set by regulations	
Life cycle impact on water concept	Evaporative use	Water which is evaporated during its use hence not immediately available for further use	ISO 2006; Jolliet et al. 2004; Frischknecht et al. 2009; Milà i Canals et al 2009; Pfister et al. 2009; ISO 2012
	Non-evaporative use	Water returned to any freshwater source after its use and available for further use	
	Water consumption	Freshwater withdrawals which are evaporated, discharged into different watersheds or the sea after use and embodied in products and waste	
	Water degradation	Water which is discharged in the same watershed after the quality of water has been altered	
	Irrigation water	The blue water consumed in agricultural activities	

4.3 Overview of Case Studies in Agri-Food Sector

4.3.1 Main Case Studies on Water Footprint Accounting Concept

Several studies have been published on the application of Water Footprint to food products.

Chapagain and Orr (2010) presented one of the first analysis of the water footprint of an industrial food product: the Nestlé's 'Bitesize Shredded Wheat'. This is to be considered a pilot study from which outcomes are considered by Hoesktra et al. (2011) in the definition of the Water Footprint methodology presented within the Water Footprint Network. The study preform the so called sustainability assessment limited to the water scarcity issue.

Ercin et al. (2011) presented a first attempt to perform Water Footprint at corporate level. The case study is relevant to an hypothetical sugar-containing carbonated beverage. The inventory used for the assessment included processes of the company and of its supply chain. These processes results to have the biggest contribution to the final water footprint. The results are based on the methodology presented by Hoekstra et al. (2011) and cover the blue, green and grey water footprint.

Another paper from Ercin et al. (2012) is focused on the water footprint of soy milk and soy burger produced from different raw materials (organic and non-organic soy) and origin (Canada, China, and France). The water footprint is represented at the level of accounting and a comparison between such products and their correspondent meat product is presented. This study moreover confirms the importance of adopting a supply chain perspective when studying water use.

Manzardo et al. (2012) has recently presented a case study where two different approaches in assessing water footprint of an organic strawberry jam pot are adopted: the one from Hoekstra et al. (2011) and the one from Ridoutt et al. (2010). In this case the water footprint accounting and scarcity assessment have been performed including all the ancillary processes and products (such as packaging transportation, etc). Such processes resulted to have a relevant contribution to final product water footprint therefore it confirms the need of adopting life cycle perspective when looking at the impacts related to water.

4.3.2 Main Case Studies on Life Cycle Impact on Water Concept

The case study from Ridoutt et al. (2010) presents the application of LCA principles to the water footprint assessment of two food products produced in Australia: Dolomio Tomato Sauce. It is one of the first attempt to go further water accounting and consider effects of water use on local water availability and therefore their contribution to local water scarcity. Results demonstrate how different products and production systems can be compared based on their contribution to water scarcity.

Further development in the LCA frameworks come from Milà i Canals et al. (2009, 2010). The goal of the study is to support a better comprehension on how to include freshwater issue within the LCA framework. In particular this method suggest how to deal with inventory compilation and suggests two midpoint indicators such as Freshwater Depletion (relevant to the availability for future generation) and ecosystems related impacts. The proposed methodology has been applied to the production of broccoli in UK and Spain.

The multiple case study presented by Jeswani and Azapagic (2011) considers the corn-derived ethanol produced in 12 countries and discusses different methods for inventory modelling and impact assessment for water use in life cycle assessment. The paper compares the impacts of freshwater consumption in the different countries: the results show a huge variation between different methods and demonstrate the need for a standardised methodology to assess the impacts of water use on a life cycle basis.

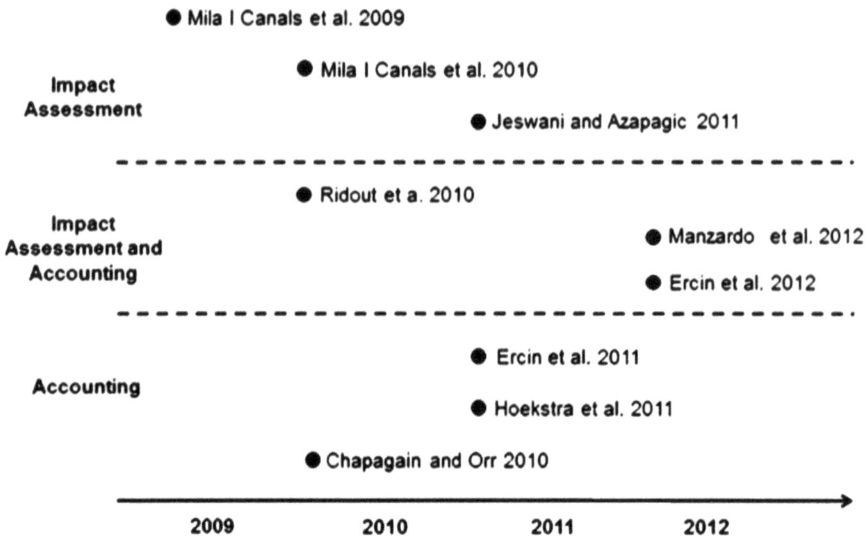

Fig. 4.1 Evolution of water footprint applications in agri-food sector

4.4 Discussion

Through the literature analysis related to case studies in agri-food sector, it is possible to trace an evolution in application of the water footprint as a tool to support environmental management (Fig. 4.1).

The case studies show that, in recent years, the focus on the water footprint has been in terms of accounting and in terms of impact assessment. In particular it shows an interest in adopting the water footprint for accounting and impact assessment tool at the same time.

Different water impact assessment methodologies developed by various authors have some limits (Table 4.2).

Referring to the virtual water concept, the main limitation is the inability of the model to quantify the impacts (since its goal is rather to quantify the water amount captured in products that can be transported). Furthermore, the measure is expressed in volume of water contained.

The water footprint accounting concept exceeds the first limit of the Virtual Water and arrives to quantify the direct and indirect consumption of water; however, only the grey water footprint component can be considered a mid-point impact assessment on water use; moreover it does not consider the environmental impacts on water resources resulting from other environmental impacts, such as air, soil, biodiversity (Scipioni et al. 2010; Jeswani and Azapagic 2011). Moreover, the water footprints are expressed on a volume basis: although the volumetric water footprint indicator is useful from the water-resource management perspective, it does not reflect the potential environmental impacts of the water use, which are important from the life cycle perspective (Jeswani and Azapagic 2011).

Table 4.2 Benefits and limits in different water use assessment concepts

Freshwater use concept	Benefits	Limits
Virtual water concept	First initiative in response to water scarcity	Inability to quantify water impacts
	First attempt to quantify water use	Measure expressed in volume of water contained
Water footprint accounting concept	Measuring impact of human activities on fresh water	Failure to consider environmental impacts on water resources resulting from other environmental impacts
	Definition of three different types of fresh water	Measure expressed in volume of water
	Direct and indirect impacts assessment of water use and consumption	Measure the impact limited at the mid-point level and related the grey water
Life cycle impact on water concept	Adoption of a life cycle approach	Lack of completeness and detail in life cycle inventory databases of water use and consumption
	Measure the impact at the mid- and end-point level	Failure of life cycle impact assessment methods to consider different input and output water types on its origin and lack of water quality parameters
	Measure including eco-factors and expressed in human health, ecosystem quality ad resources	Lack of site-specific and seasonal references in impacts assessment and strong dependence of results on impact assessment method

Life cycle impact assessment on water concept presents the main limits of LCA methodology. Particularly, water use is not systematically recorded in life cycle inventory databases and the life cycle impact assessment methods do not consider different input and output of freshwater flow according to water types and its origin (Koehler and Aoustin 2008; Kounina et al. 2013). This is probably because LCA is largely time and location independent, while the impacts and issues related to water are seasonal and significant at local level within the confines of the watersheds and river basins. Furthermore, the available impact assessment methods do not have a set of quality parameters (Kounina et al. 2013). Finally, assessment results depend strongly on the assessment method: this compromises the solidity of the LCA studies and demonstrates the need for a standardized methodology for assessing the impacts of water use on a life cycle basis (Jeswani and Azapagic 2011).

4.5 Future perspectives

To tackle this major environmental concern, various initiatives were recently launched in order to develop and standardize analytical tools to measure and assess freshwater use at regional and global scale and to improve the overall management of freshwater resources as well as the overall environmental performance of

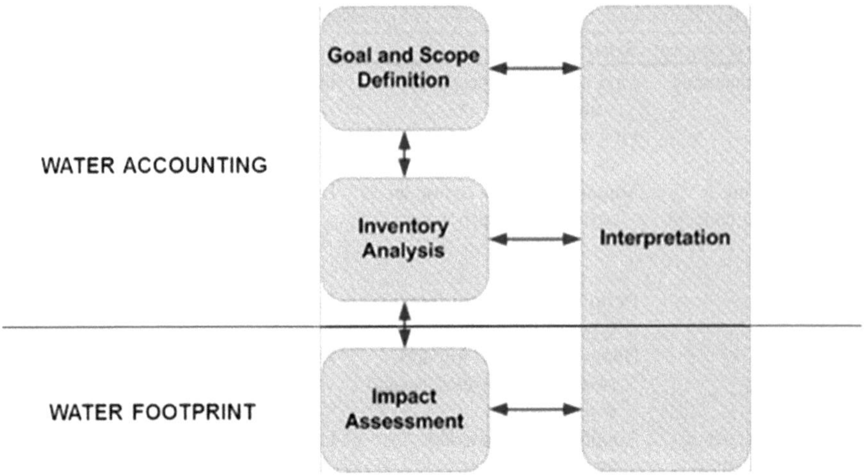

Fig. 4.2 Different steps and different levels of Water footprint framework in ISO 14046 Committee Draft. (Adapted from ISO 2012)

products and operations. Among these initiatives is the Water Footprint Network (WFN), international circuit with the participation of public and private stakeholders in order to promote discussion and comparison of water impact assessment using the water footprint accounting (Hoekstra et al. 2011). Also the World Business Council for Sustainable Development (WBCSD) and the Society of Environmental Toxicology and Chemistry (SETAC) promote initiatives also aim to set up public-private partnerships to assist companies in the implementation of water sustainability policies (WBCSD 2010; Koehler and Aoustin 2008).

The International Organization for Standardization (ISO) since 2009 established a specific working group (ISO/TC207/SC5/WG8) in order to determine an international standard on water footprint. The standardization process should be completed in 2014 with the publication of ISO 14046: "Water Footprint. Principles, requirements and guidelines". This process build its content on all the experiences lead by the scientific community related to water impact assessment (ISO 2012).

Even if the ISO working group is still working on the detailed requirements, it is already possible to point out the main concepts on which the standardized methodology is based.

First of all water footprint is defined as metric(s) that quantify(ies) the potential environmental impacts related to water, then it is possible to compare water footprint with other impact assessment indexes. Moreover, the methodology can be applied to product, processes and organization, therefore its use may be at the level of accounting and also management. Secondly, focusing on the structure of the water footprint, it is the same of the LCA study, in fact there are four steps (goal and scope definition, inventory analysis, impact assessment and interpretation), and it clearly differs from what is to be considered water accounting, related to the inventory analysis, and water footprint, determined at impact assessment level (Fig. 4.2). Furthermore, the inclusion/exclusion of data and processes shall be based on the life cycle approach as described by the LCA standards, then the ISO 14046 can be used in integration with the LCA standard ISO 14040. Another important issue of such a methodology is the concept of comprehensiveness adapted from LCA standards: the study shall consider all relevant impacts related to water, also including water impacts derived indirectly from other environmental impacts.

References

Allan JA (1998) Virtual water: a strategic resource, global solutions to regional deficits. Groundwater 36(4):545–546

CAWMA (2007) Water for food, water for life: a comprehensive assessment of water management in agriculture. Earthscan, London

Chapagain AK, Orr S (2010) Water footprint of Nestlé's 'Bitesize Shredded Wheat'. Water Footprint pilot project. Water Footprint Network, Geneva

EPA (2006) Life cycle assessment: principle and practice. National Risk Management Research Laboratory—Office of Research and Development, U. S. Environmental Protection Agency, Cincinnati

Ercin A, Aldaya MM, Hoekstra AY (2011) Corporate water footprint accounting and impact assessment: the case of the water footprint of a sugar-containing carbonated beverage. Wat Res Man 25(2):721–741

Ercin A, Aldaya MM, Hoekstra AY (2012) The water footprint of soy milk and soy burger and equivalent animal products. Ecol Ind 18:392–402

EU (2012) Blueprint to safeguard Europe's water resources. European Union, Brussels

Frischknecht R, Steiner R, Jungbluth N (2009) The ecological scarcity method—Eco-factors 2006. A method for impact assessment in LCA. Federal Office for the Environment, Bern

Hoekstra AY (2010) The water footprint: water in the supply chain. Environmentalist 93:12–13

Hoekstra AY, Chapagain AK (2008) Globalization of water: sharing the planet's freshwater resources. Blackwell, Oxford

Hoekstra AY, Chapagain AK, Aldaya MM, Mekonnen MM (2011) The water footprint assessment manual: setting the global standard. Water Footprint Network, Enschede

Hunt RG, Franklin WE (1996) How it came about. Personal reflection on the origin and development of LCA in the USA. Int J Life Cycle Ass 1(1):4–7

IPCC (2008) Climate change and water, technical paper VI. Intergovernmental Panel on Climate Change, Geneva

ISO (2006) ISO 14040—Environmental management—Life cycle assessment —Principle and framework. International Organization for Standardization, Geneva

ISO (2012) ISO/CD 14046 Life cycle assessment—Water footprint—Requirements and guidelines. International Organization for Standardization, Geneva

Jeswani HK, Azapagic A (2011) Water footprint: methodologies and a case study for assessing the impacts of water use. J Clean Prod 19:1288–1299

Jolliet O, Müller-Wenk R, Bare J, Brent A, Goedkoop M, Heijungs R, Itsubo N, Pena C, Pennington D, Potting J, Rebitzer G, Stewart M, de Haes H, Weidema B (2004) The LCIA midpoint-damage framework of the UNEP/SETAC life cycle initiative. Int J Life Cycle Assess 9(6):394–404

Koehler A, Aoustin E (2008) UNEP/SETAC Life Cycle Initiative SETAC Europe Annual Meeting, Project group on assessment of use and depletion of water resources within LCA. Warsaw

Kounina A, Margni M, Bayart JB, Boulay AM, Berger M, Bulle C, Frischknech R, Koehler A, Milà i Canals L, Motoshita M, Núñez M, Peters G, Pfister S, Ridoutt B, van Zelm R, Verones F, Humbert S (2013) Review of methods addressing freshwater use in life cycle inventory and impact assessment. Int J Life Cycle Ass 18:707–721

Manzardo A, Mazzi A, Niero M, Toniolo S, Scipioni A (2012) Water footprint accounting of organic and non-organic strawberries including ancillary materials: a case study. In: Corson MS, van der Werf HMG (eds) Proceedings of 8th international conference on life cycle assessment in the agri-food sector. (LCA Food 2012). INRA, Rennes, pp 45–50

Milà i Canals L, Chenoweth J, Chapagain A, Orr S, Antón A, Clift R (2009) Assessing freshwater use impacts in LCA: part I—inventory modelling and characterisation factors for the main impact pathways. Int J Life Cycle Ass 14(1):28–42

Milà i Canals L, Chenoweth J, Chapagain A, Orr S, Antón A, Clift R (2010) Assessing freshwater use impacts in LCA, part 2– case study of broccoli production in the UK and Spain. Int J Life Cycle Ass 15(6):598–607

PF (2010) Loi n.2010–788 du 12 juillet 2010 portant engagement national pour l'environnement "Grenelle II". NOR: DEVX0822225 L. Parlement Français. Paris

Pfister S, Koehler A, Hellweg S (2009) Assessing the environmental impacts of freshwater consumption in LCA. Environ Sci Tech 43(11):4098–4104

Ridoutt BG, Juliano P, Sanguansri P, Sellahewa J (2010) The water footprint of food waste: case study of fresh mango in Australia. J Clean Prod 18(16–17):1714–1721

SETAC (1991) A technical framework for life cycle assessment. Workshop report from the Smuggler Notch, Vermont. SETAC, Brussels

Scipioni A, Mazzi A, Manzardo A (2010) La gestione strategica della risorsa idrica, gli strumenti per valutare e comunicare il water foortprint. Qualità (6) 8–10

UNESCO (2009) The United Nations world water development report 3—water in a changing world. United Nations Educational Scientific and Cultural Organization, Paris

WBCSD (2006) Business in the world of water: WBCSD water scenarios to 2025. World Business Council for Sustainable Development, Geneva

WBCSD (2009) Water for business, initiatives guiding sustainable water management in the private sector. World Business Council for Sustainable Development, Geneva

WBCSD (2010) The global water tool. World Business Council for Sustainable Development, Geneva

WWAP (2009) The United Nations world water development report 3: water in a changing world. UNESCO Publishing, Paris

Chapter 5
Marketing Strategies for the Agri-Food Sector: Adaptation Requirements and Contradictions

Grazia Calabrò

Abstract The hotly debated topics of global warming and climate change cut across all sectors of human activity, influencing policy decisions and modifying consumer habits and lifestyles. The food industry also makes a significant contribution to climate change in terms of greenhouse gas emissions, at every stage of the production process. Indeed, there is no doubt that agriculture and climate change are characterized by a complex cause-effect relationship. This chapter analyses the effects of the hoped for reorganization of the agri-food sector in with a view to sustainability, underlining that this must regard three main considerations: the need to blend the dual production and conservation functions of agriculture; the need to create consumer lifestyles that make dietary and environmental well-being compatible; the identification of appropriate certification and consumer information tools which highlight the measures undertaken in order to limit the effects on climate along the whole food production chain.

Keywords Climate change · Food industry · Consumers' habits · Consumers' information · Low carbon economy · Carbon footprint · Labeling

5.1 Introduction

Climate change is an unresolved problem that cuts across all sectors of human activity and which the global community needs to face up to. In this context, the relationship between agriculture and environment has also become a subject of discussion and debate in relation to sustainability policies.

The arguments supporting the relationship between agriculture and climate change concentrate, on the one hand, on a reorganization of the sector in terms of sustainability and, on the other hand, on the possible repercussions that this reorganization may have on food safety. Indeed, there is no doubt that agriculture and climate change are characterized by a complex cause-effect relationship.

G. Calabrò (✉)
Department SEAM, University of Messina, Piazza S. Pugliatti 1, 98122, Messina, Italy
e-mail: grazia.calabro@unime.it

R. Salomone, G. Saija (eds.), *Pathways to Environmental Sustainability*,
DOI 10.1007/978-3-319-03826-1_5,

Agriculture is responsible for a modest amount of carbon dioxide emissions, which are the main contributing factor to climate change; the greatest effect agriculture has on climate change is caused by emissions of nitrous oxide and methane resulting from certain important activities. It is forecast that, without corrective measures, the contribution of agriculture to climate change, as regards nitrous oxide and methane emissions, will increase significantly.

The reorganization of the sector in terms of sustainability means that increased production must be compatible with the requirement to safeguard and protect resources in order to avoid food safety risks, in the sense of physical and economic access to sufficient, safe and nutritious food.

Changes in general climatic conditions and the resulting variability of water supplies could have repercussions on both the quantity and quality of food products.

As well as performing strategic economic and social tasks, as part of its new multifunctional role agriculture also has the task of protecting and conserving the rural environment since, in the same way as other sectors of the economy, it causes potentially prejudicial environmental effects.

It goes without saying that, on an operational level, two situations come about: on the one hand agriculture, in carrying out its production function, is subject to ever tighter environmental restrictions aimed at moving this sector away from an intensive production model with high environmental impact and which may bring about a decline in production, with social and environmental implications, as well as economic ones; on the other hand, the subsidies paid out to agriculture tend to limit in some way the losses resulting from environmental restrictions, but are not perfectly capable of giving rise to new production models oriented towards respect for the environment, almost creating a conflict between the production and conservation roles of agriculture.

It is, therefore, evident that the food sector, which is closely tied to agriculture and highly dependent on it, also needs to be reorganized with a view to reestablishing, upstream, a situation of balance, rather than conflict, between agriculture and environment and to actively involving, downstream, all the actors in the chain, particularly consumers, who are in a position to move the market through their consumption choices, but who can also be moved by the market towards purchasing behavior and dietary habits which represent a better response to the problem.

Evaluation of these aspects must therefore be organized bearing in mind the multiple factors involved in the relationship between climate change and the food industry, which has characteristics (long production chain, outsourcing of numerous phases, involvement of large number of actors, etc.) that make it difficult to manage the numerous impacts to be taken into consideration when measuring the contribution of the food industry to climate change and which highlight the need for innovation in agri-food management and practices, with a view to integrating food security, food safety and food sustainability.

5.2 Relationship Between the Food Industry and the Environment

The contribution of agriculture to the production of greenhouse gases worldwide has increased over the years and future scenarios, envisaged by the International Panel on Climate Change (IPPC), indicated the risk that by 2030, there would be an increase of 35–60 % in nitrous oxide and of 60 % in methane released by the agricultural sector.

The activity generating the greatest amount of greenhouse gases is deforestation, followed by land fertilization and emissions of gases coming from cattle digestion (Cline 2007).

The potential repercussions envisaged as a result of climate change mainly involve:

- the management of water resources: it is expected that there will be an unequal distribution of water resources as a consequence of the decline in resources in arid or semi-arid areas and an increase of the overall amount in the northern hemisphere;
- the spread of disease and contamination in agricultural products and foodstuffs as a consequence of extreme events such as floods and hurricanes, higher temperatures and more intense rainfalls;
- the scarcity of food resources and impacts on social security.

The contribution of the food sector to climate change is an element of every stage of production; it is estimated, for example, that 49 % of greenhouse gas emissions can be attributed to the agricultural phase, about 11 % to the production phase and 18 % to final consumption. Moreover, the need for maximum food safety, from a hygiene-health viewpoint, involves energy intensive actions and procedures, such as refrigeration, which, according to estimates made by the International Institute of Refrigeration, accounts for about 15 % of energy consumption worldwide (James and James 2010) (IPPC 2007).

The need to move towards sustainable food consumption cannot avoid the consideration that food is a basic necessity and cannot be substituted by other services and that consumer choices cannot be rigidly linked solely to the need to reduce environmental impacts, given that these are strongly influenced by organoleptic acceptability, social and cultural aspects and reflect social and customary changes. For this reason, in order to rethink consumption habits and move consumers towards more sustainable dietary choices, we need to concentrate on both quantitative and qualitative aspects (the type of food and its composition) and on evaluation of market opportunities.

5.3 Actions Undertaken by the Food Production Chain

The intensification of trade, changes in society, development in technology and transport have all provided a strong push for product and process innovation in the agri-food sector and have contributed to the strong economic and employment performance of the allied industries in this sector.

In particular, the development process has allowed production to be de-territorialized and guaranteed an increase in food supplies both in spatial and temporal terms. Indeed, large scale distribution is in a position to place enormous quantities of food onto the market; this food comes from all over the world and reaches its various destinations completely safely.

These aspects are the strong points of the food industry and have led to a growth in consumption and the development of various forms of product differentiation; however, at the same time, there is a negative side to this in terms of sustainability. Indeed, while in the past the food sector and technological innovation concentrated their attention on facilitating product access and market availability, nowadays the concept of sustainability means there is need for a rethink, or perhaps even a complete shake up of traditional development and consumption models, and it becomes the new element on which to base marketing.

The problems of climate change and its relationship with the agri-food sector need to be managed while bearing in mind various aspects:

1. *The need to blend the dual production and conservation functions of agriculture*: The impact of agriculture in the territory and, in particular, on soil, following intensive land use, creates the risk of accelerating erosion and reducing organic substance which is indispensable for the development of all activities dependent on agriculture. Consequently, the development of agricultural practices that tend to reconcile productivity and soil protection are based on conservative agriculture, which requires the abandonment of conventional growing techniques and the adoption of reduced land use or, in extreme cases, land disuse. It should be noted, however, that the conservation function must not come into conflict with the production function, given that the latter provides agriculture with its primary role of supplying goods for the production of foodstuffs, but also in terms of collective interest. Indeed, agriculture plays an important role in terms of territorial enhancement and protection. Soil impoverishment is the root cause of the main hydro-geological instabilities, which are more and more frequent, and is linked, on the one hand, to the decline in agricultural activities and, on the other hand, as has been pointed out, may be caused by agriculture itself (Vieri 2012a). The implications are negative from both a social and economic viewpoint and in terms of access to food supplies, given that production may be lost due to structural instability and climatic consequences. Thus, agriculture performs a function of territorial protection and actions need to be aimed at avoiding the abandonment of agricultural practices, also through subsidies, and at encouraging the modernization of agricultural production structures and providing incentives for sustainable agronomic practices (Vieri 2012b). In this way action can

be taken on reducing methane and nitrous oxide emissions and, at the same time, the agricultural sector maintains its essential role in terms of absorption of carbon dioxide, the main greenhouse gas.

2. *The identification of appropriate consumer information tools for orienting choices*: The introduction of communication tools for the purposes of rewarding those who pursue sustainability may transform the problem of climate change from a restriction to a market opportunity.
   Indeed, we need to consider that consumers are worried about the consequences of climate change but do not realize the considerable importance of their daily purchasing habits in terms of contributing to the problem, and therefore do not change them (Gallup Organization 2009).
   It goes without saying that consumers continue to devolve decisions on both the causes and actions to be taken to agents belonging to specific sectors, mainly industrial ones, and not to their consumption choices. For example, interest in organic agriculture is more closely tied to the desire to eat "healthy and natural" products than to careful consideration of environmental aspects; in this sense, the greater attention paid to agricultural practices is more closely linked to better perception of food safety issues than to any environmental benefits.
   Taking advantage of the fact that consumers tend to link environmental issues with their personal interests and that this influences their purchasing choices, numerous strategies have attempted to generate a latent need for "climate change" and to supply, at times in a debatable way, what seems to be the most appropriate answer to the problem; this answer may often be more psychological than practical. For example, although the food industry causes environmental impacts at in every phase of production, from the field to the table, consumers seem to more easily perceive the role of the food product transportation phase. The effects of food product transportation on the environment have now taken hold so much in the mind of consumers that a short chain and the concepts of *food miles* and *friendlier miles* have become an integral part of consumer choices and various systems of climate labeling have been introduced onto an increasing number of food products on the market.
   Climate labeling systems, above all if supported by third party certification, are an effective marketing tool which informs consumers of the measures undertaken in order to limit effects on the climate along the production chain and guarantees the reliability of the information provided. Indeed, it has been shown that consumers alter their purchasing choices when they are sure that their choice will make a clear, measurable and tangible contribution to the environment.
   It needs to be remembered that the market penetration of this type of tool is not geographically homogenous and is conditioned by varying degrees of environmental sensitivity, meaning that in some markets it satisfies an explicit need, while in others it induces a still strongly latent element, thus attractive from a marketing viewpoint.
   The objective is that of making sure consumers recognize the information on CO_2 emissions during transportation as an added value, incorporating it into the traditional quality and price parameters.

Thus, on the basis of the equation constructed by consumers, a short chain does not mean lowering economic and social costs, but rather choosing local products, which is seen positively as an expression of a safer and better quality product. In this way, promotion of the concept of *food miles* among consumers may serve as a driving force for developing the products of those countries, such as Italy, in which people's ties to their origins are very strong and tend to be valued (Barilla Center for Food and Nutrition 2009).

It is clear, however, that these concepts actually supply consumers with partial information on environmental impact, given that considering only the distance in kilometers is rather restrictive; thus they can be criticized from the viewpoint of accurate information, but they are effective from a marketing point of view (Schimidt 2009).

Thus, other methodologies become more appropriate, such as *carbon footprint labeling or climate labeling,* which are now to be found on the market and which are aimed at supplying more objective and complete information on the efforts made by companies to reduce the "ecological footprint" of their products (Atsushi et al. 2010). It is worth noting that interest in calculating carbon footprints has involved new feasibility studies for well established and widely known ecological labels, such as the German ecological label Blue Angel, to which further specifications have been added with regard to greenhouse gas emissions during the life cycle, and the European Ecolabel, for which applicability to food products is being studied; this study, is currently somewhat puzzling due to the possible confusion and overlaps which could mislead consumers with regard to product characteristics (Finkbeiner 2009) (Pathak et al. 2010). Various initiatives have also been undertaken on an international level in Japan, the United States, Taiwan and Korea.

All these initiatives are characterized, however, by a lack of coordination and uniformity in calculation procedures, which represents a weak link in the communication chain. For this reason, there are coordination actions underway both on a European and international level with the objective of achieving a single system appropriate for market purposes (Calabrò and Lagioia 2011).

3. *The need to create consumption lifestyles which balance dietary and environmental well-being*: Social and economic wealth and the widespread availability of food on the market have encouraged, on the one hand, an increase in consumption and a change in dietary habits but, on the other hand, the need to deal with environmental issues that may have important implications on a social level. Models of consumption have always been representative of culture and of the characteristics of the social context in which they developed and, thus, need to be updated and modified on the basis of the new requirements that arise.

The dietary choices of individuals and households can have an increasing effect on the ecological balance of the planet and should be considered in the path towards sustainability. A change in diet is therefore one of the most important proposals for achieving a sustainable lifestyle. Future food consumption patterns should reflect a growing consideration of the state of the environment (Carlsson-Kanyama 1998)

Today, higher calorie intakes, resulting from the quantity, variety and novelty of food consumption, associated with a stressful and sedentary lifestyle, have, in the long term, led to an alarming growth in cardiovascular diseases, diabetes and obesity; these are degenerative diseases which have economic as well as social consequences, especially in terms of health care spending. Indeed, eating more and badly negatively influences not only people's health and psychophysical well-being, but we could also mention the environment.

The food industry has taken on board the concerns expressed by nutrition and medical science regarding the food-health relationship and now markets products with a high service content, satisfying the requirements imposed by new food models and meeting the needs of consumers, who are reluctant to change their dietary habits in line with new guidance and prefer to turn to the healthy versions of industrial foods. Purchasing choices therefore, reflect the new tendency to satisfy first and foremost needs of health and well-being, needs which have been induced through seductive marketing strategies (Calabrò and Postorino 2010).

Food risk is thus no longer perceived as tightly correlated to the intrinsic nature of products, but is extended to a wider level on which food risk derives from an incorrect and unbalanced combination of food. Consequently, rather than varying their diet and moving away from unbalanced diets with high levels of fat, meat and sugars, consumers carefully read the signals sent out by the food industry, which takes on the inappropriate role of dietary education.

The policy of food education is highlighted as a means of prevention but is exploited by the industry, which continually sends messages emphasizing the relationship between diet and health, inducing consumers to concentrate on this aspect and to find effective consolation for their choice of *convenience foods*, for which they are even prepared to pay extra due to the quality of use that these foods incorporate.

Moreover, health risk concerns become an integral part of the food industry's environmental strategy from two points of view:

- Firstly because it has been shown that consumption of products with high fat and sugar content is not only problematic as regards the diet-health relationship, but also as regards the environment, given that these products are associated with higher environmental impacts. Thus, promoting diets with high levels of carbohydrate, fruit and vegetable consumption proves to be more balanced not only from a nutritional viewpoint, and thus also from a viewpoint of health, but also more *low carbon*, and, thus, a key point of the green economy;
- Secondly because by extending the relationship of food products with health to underlying environmental aspects as well, it is possible to use the consideration that protecting health benefits the environment and protecting the environment benefits health as an effective marketing tool.

5.4 Conclusions

Food consumption directly or indirectly causes emissions of both energy related greenhouse gases and greenhouse gas emissions from other sources.

In future the food industry will find itself having to face up to the growing problem of sustainability in various ways, such as the following:

- Upstream, by adopting a new vision of agriculture, which can only continue to perform its production function in harmony with its conservation function. These aspects need to be coordinated, otherwise acting environmentally from a production viewpoint negatively influences the conservation and social aspects. De-territorializing or de-seasoning production influences economic aspects but also has important environmental effects and considerable social implications because the price competition mechanism could lead to a decline in productivity and profitability for local operators, who could be induced into abandoning their fields; besides, if the areas under cultivation are reduced, then the strategic function of agriculture in terms of soil protection disappears. Supporting the idea that the type of agriculture capable of ensuring protection of the environment is one that does not produce but rather performs conservation tasks has ended up weakening agriculture itself, with its traditional role as a sector of production in decline and without its multifunctional role being fully recognized.
- Downstream, by involving all the actors in the production chain in two aspects: communication and thus use of environmental certification tools alongside traditional forms of information used to orientate consumer choice; education in order to move the method of choice away from how to eat sustainably towards what to eat sustainably. This element is especially important since without correct education no results can be achieved, given that increased income results in increased demand for luxury products (such as meat). Paradoxically the economic crisis, which has contracted incomes, has moved consumers towards low carbon consumption, demonstrating once again that the price mechanism is decisive in influencing choice.

In conclusion, the whole food production chain needs to follow a common pathway, given that the resulting effects are all interconnected and interdependent.

References

Atsushi K, Yuji W, Akiko K, Tomohiro M, Masako O, Sho-ichi G, Daisuke T, Dongsheng C, Takashi O, Ippeida D (2010) Interactive effects on carbon footprint information and its accessibility on value and subjective qualities of food products. Appetite 55:271–278

Barilla Center for Food & Nutrition (2009) Cambiamento climatico, agricoltura e alimentazione. http://www.barillacfn.com/uploads/file/62/1244800592_ClimateChangeIT_BarillaCFN_0609.pdf. Accessed 12 Dec 2012

Calabrò G, Lagioia G (2011) L'etichetta climatica sui prodotti alimentari: un nuovo approccio dell'industria alimentare. In: Atti del XXV Convegno Nazionale di Merceologia Contributo delle scienze merceologiche per un mondo sostenibile. Forum (eds), Trieste

Calabrò G, Postorino E (2010). I claims nutrizionali e salutistici: strumento di informazione merceologica, di tutela del consumatore e di marketing. In: Percivale F, Varese E, Arese M (eds) Ambiente, Internazionalizzazione, Sistemi, Merci, Energia. Celid, Torino

Carlsson-Kanyama A (1998) Climate changes and dietary choices—how can emissions of greenhouse gases from food consumption be reduced? Food Policy 23:277–293

Cline W (2007) Global Warming and Agriculture. Centre for Global Development, USA

Finkbeiner M (2009) Carbon footprinting—opportunities and threats. Int J Life Cycle Assess 14:91–94

Gallup Organization (2009) Europeans' attitudes towards the issue of sustainable consumption and production Analytical report. http://ec.europa.eu/public_opinion/flash/fl_256_en.pdf. Accessed 18 Dec 2012

IPCC (2007) Fourth Assessment Report of the Intergovernmental Panel on Climate Change. Cambridge University Press, Cambridge

James SJ, James C (2010) The food cold-chain and climate change. Food Res Int 43:1944–1956

Pathak H, Jain N, Bhatia A, Patel J, Aggarwal PK (2010) Carbon footprints of Indian food items. Agr Ecosys Environ 139:66–73

Schmidt HJ (2009) Carbon footprinting, labelling and life cycle Assessment. Int J Life Cycle Assess 14(Suppl 1): 6–9

Vieri S (2012a) Agricoltura settore multifunzionale allo sviluppo. Edagricole, Milano

Vieri S (2012b) The common agricultural policy and measures for the protection and conservation of the environment: contrasts and balances and new development models for the future. Int J Env H 6(1):48–62

Chapter 6
Hypercritical Separation Technology (HYST): A Sustainable Technology for Agricultural By-products Valorization

Pierpaolo Dell'Omo, Francesca Luciani, Raffaella Preti and Giuliana Vinci

Abstract Agricultural activities produce worldwide about 3 billion tonnes of by-products and residues per year, that represent an important source of food, feed and bioenergy. The technologies currently available for these resources exploitation are not economically advantageous and not environmental friendly. Hypercritical Separation Technology (HYST) is an innovative technology based only on physical process, for the disaggregation of biomass. By this technology, a flour for human consumption with high vitamin and minerals content can be obtained from cereal bran. The HYST system has proved to be efficient to produce also feed with high nutritional value and improved digestibility. In the bioenergy sector, this technology could play a crucial role, for a sustainable and cheap production of second generation biomethane. Future projects to explore the potentialities of this technology will involve new agricultural residues, such as grape pomace, source of antioxidants and rice bran, source of proteins, regarding the food application, and the production of chemicals from fermentative process of lignocelluloses biomass for green chemistry applications.

Keywords Waste biomass · Functional foods · Biofuels · Sustainable technology · Green chemistry

6.1 Introduction

The agricultural sector produces a large amount of by-products and wastes, representing an increasing interest as industrial crops both due to economic reasons and environmental concerns. Agricultural activities produce worldwide about 3 billion tonnes of by-products and residues per year, that could represent an important

R. Preti (✉) · G. Vinci
Department of Management, Sapienza University of Rome, Rome, Italy
e-mail: raffaella.preti@uniroma1.it

P. Dell'Omo
DIAEE, Sapienza University of Rome, Rome, Italy

F. Luciani
CRIVIB, Istituto Superiore di Sanità, Rome, Italy

R. Salomone, G. Saija (eds.), *Pathways to Environmental Sustainability*,
DOI 10.1007/978-3-319-03826-1_6,

source of food, feed and raw materials for the production of second generation biofuels and green chemistry products. In Italy 20 million tonnes of agricultural residue every year rot in the fields (ENEA 2006). In particular the world milling industry transforms every year about 350 million tonnes of wheat, producing about 80 million tonnes of by-products (about 25 %), with no real value, end up totally in feed, carrying starch, protein, over 70 % of the vitamin B6 contained in the grain, more than 50 % of the B5 more than 33 % of vitamins B1 and most of Fe, Zn, Mg, K, that remain caught in a web of indigestible fiber. These nutrients are an important source of food, even more important if directed to support programs for food in developing countries. Other by-products, such as grape residue from the processing of the grapes retain proteins and lipids of good nutritional value, as well as phenolic compounds that are extremely interesting for their beneficial effects on human health (Ruberto et al. 2007).

Currently, the food industry shows a growing interest in the production of functional foods, which are generally obtained by adding "bioactive" ingredients. The availability of technologies capable of producing functional foods directly from biomass containing these bioactive compounds by removal of indigestible substances and antinutritional factors, could be a turning point for this area.

Even more, the EU directive Waste Framework Directive 2008/98/EC (EU 2008), as regards food waste, gives priority to the reduction of waste at source, followed by reuse, recycling and recovery, with the elimination as a last choice.

The extraction of nutrients from residues of the food industry with the use of technology currently available is not, however, a process economically advantageous, because of low yields in the face of production processes extremely expensive. More than 20 million tonnes of straw and prunings, in codigestion with animal wastes, can potentially meet about 10 % of national demand for natural gas, which we are importing almost entirely from overseas. The Directive 2009/28/EC (EU 2009) on the promotion of energy from renewable sources establishes a 10 % target of energy needs in the transport sector to be covered by renewable fuels by 2020. In 2012 Italy will reach its 4.5 % share, which will almost entirely be covered by biofuels produced outside Italy.

The limiting factor for an effective utilization of lignocellulose in the processes of conversion into biofuels is the difficulty of hydrolysis of cellulose and hemicellulose, strongly linked with the matrix of non degradable lignin. Efficient technologies of pretreatment, with low operating costs and available at industrial level are therefore indispensable to facilitate the enzymatic hydrolysis and to provide an adequate production of second generation biofuels.

6.2 Technology

The Hyst technology (Hypercritical Separation Technology—Patent Application WO 2011/061595 A1), invented by Eng. Umberto Manola, consists of a set of machines diagrammed to process primarily biomass, but also inorganic substances with a water content < 15 %, exclusively through a physical process.

Fig. 6.1 HYST plant for the processing of agricultural residues

The system causes the disaggregation of the plant structure through reciprocal collisions between the particles within a current of air, without the aid of grinding rolls. In this way the raw material remains at room temperature, thus preserving its nutritional and organoleptic properties. The product of the disaggregation is then separated into several fractions of different physical and chemical characteristics, which can be used individually or mixed.

The system (Fig. 6.1), which constitute self-sufficient units, is modular; by varying the unit number is then possible to adjust the production capacity to the availability of biomass and/or the demands of the market.

The HYST plant has a modest energy demand, about 20 kWh for biomass ton treated, in comparison to the 500–1000 kWh of the other systems of treatment, with no need of chemical substances or water, with zero emissions and low operating costs.

Thanks to this technology from 100 kg of biomass is possible to extract 20 kg of lignin and cellulose and 25 of digestible fiber for zootechny. The remaining can be treated to produce 27 l of ethanol.

6.3 Applications

Encouraging preliminary studies have been carried out on the application of HYST process in several strategic sectors, hereafter are described some of the most relevant.

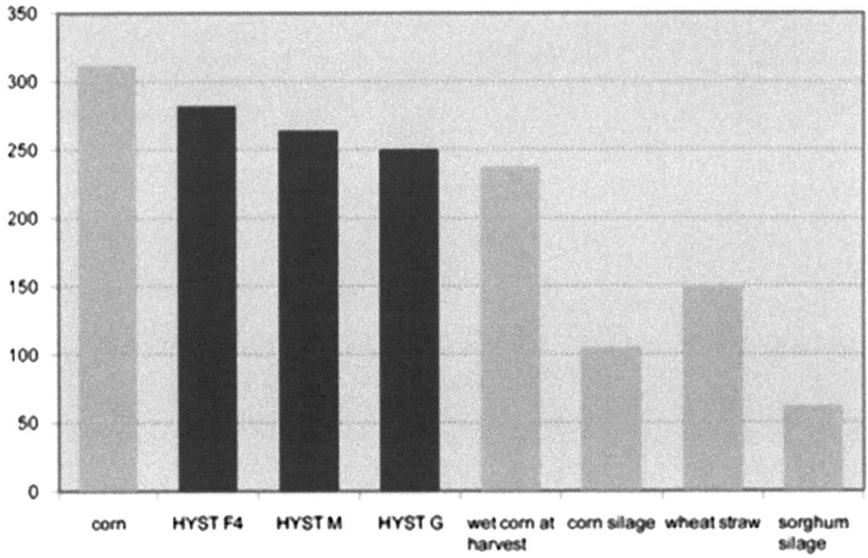

Fig. 6.2 Biomethane production (m^3/t w.b.) of HYST matrices from cereal straw compared with production of other biomass

6.3.1 *Green Energy*

The Italian transposition of the European Directive 2009/28/EC (EU 2009) for the renewable energies has established an increase of five times of the national biogas production. From the other side the ethical debate on the subtraction of land for the no-food production is always present. To achieve EU goals it will therefore be necessary to use second-generation biofuels (i.e. produced from agricultural by-products and residues), currently absent from the market because production technology still needs to be fully developed.

The HYST technology applied on straw flour has resulted in an energy density, in terms of methane production per ton of raw material, increased up to three times compared to corn silage (which is the reference biocrop; Fig. 6.2).

The HYST biomethane would be better valorized not in the electricity sector but in the automotive sector, where it could cover about 9 % of energy needs in the Italian transport sector (based on data processed by Unione Petrolifera databook 2012); complying with *double counting* rules (*Double counting*: introduced with the Directive 2009/28/EC (EU 2009), conventionally doubles the energy content of second-generation fuels, precisely in order to encourage industry efforts in this area) near the 10 % renewable energy share target for 2020 by means of only Italian unused raw materials. The excellent results and extremely low energy consumption of HYST pretreatment allows significantly lower biomethane production costs compared to those of biofuels currently on the market. HYST biomethane can even

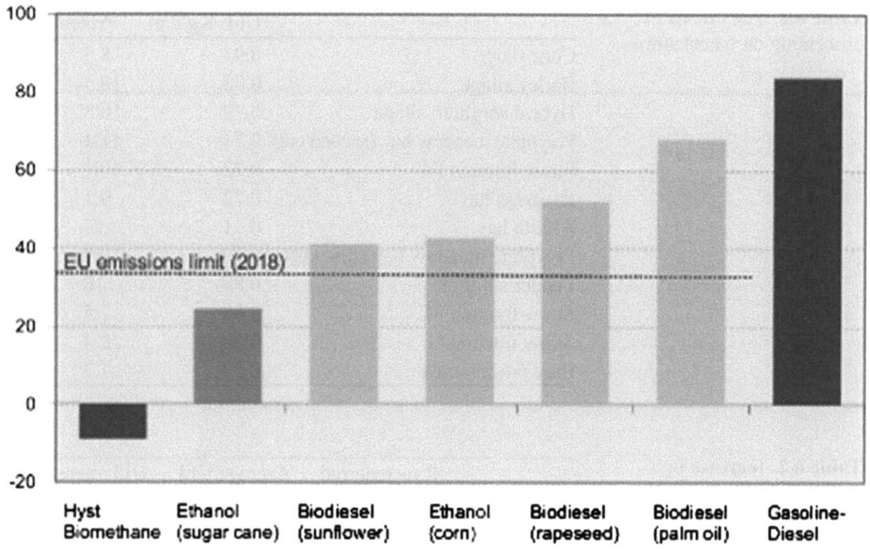

Fig. 6.3 GHG emissions [gCO_2eq/MJ] grams of CO_2 equivalent per unit of energy (MJ) released by the combustion of fuel

compete with costs of traditional fuels, 0.54 € per liter of gasoline equivalent, versus 0.67 € for first generation biomethane.

Furthermore HYST biomethane produced from residues of agricultural activities in codigestion with animal waste will not only easily meet European obligations but it will even absorb CO_2 from the atmosphere (-9 gCO_2eq/MJ; Fig. 6.3).

These results are possible because the HYST system can process agricultural residues with minimal energy consumption, along with significant environmental benefits deriving from improved management of animal wastes.

6.3.2 *Animal Feeding*

The use of agricultural and agroindustrial by-products in animal feed is an ancient practice, as it is the best way to enhance products considered as waste. Cereal straw, corn stalks or wheat bran are characterized by nutritional value (measured in UFL, forage unit for milk production) lower than citrus or beet pulps. They are, however, fundamental in the feeding of dairy cattle, beef and pigs.

The digestibility of a dry matter for animal feeding is inversely proportional to the lignin content in the matrix (Kamalak et al. 2004). To release the nutrients from lignin to increase their availability, the residues in the past were processed with chemicals or with mechanical treatments. HYST technology can operate the same disaggregation only with a physical process.

Table 6.1 The effects of processing on wheat straw

	UFL/Kg d.m.	Protein (%)
Corn silage	0.93	8.5
Barley silage	0.78	10.5
Hybrid sorghum silage	0.77	10.5
Polyphite meadow hay (second cut)	0.74	11.4
Straw fraction F4	0.72	10.1
Ryegrass hay	0.72	9.1
Alfalfa hay	0.71	17
Polyphite meadow hay (first cut)	0.70	10.9
Fodder sorghum	0.88	10
Straw fraction M	0.58	7.5
Straw fraction G	0.58	6.4
Base wheat straw	0.57	7.3

Table 6.2 Increase in nutritional value resulting from HYST treatment of three fibrous foods

	Raw material (UFL)	Average UFL increase (%)	UFL increase fraction F4 (%)
Wheat straw	0.57	6.7	25
Corn straw	0.48	6.5	33
Wheat bran	0.89	4.1	20

The laboratories of the Department of Animal Sciences of the University of Milan have studied cereal and corn straw, and wheat bran treated with HYST technology to assess the changes in nutritional value by the method of Menke and Steingass (1988). The separation process allows to obtain three fractions in order of decreasing granulometry: Gross, Medium, Fine (G, M, F4). On these samples nutritional value and digestibility analysis have been carried out. The chemical analysis showed a reduction of lignin and of the fiber fractions, with a significant increase of nutrients, especially proteins and starch, higher in the finest fractions (Dell'Omo et al. 2011).

The biological analysis showed that after lignin removal the digestibility increased substantially, in particular the F4 fraction from bran was 69.4 % in comparison of 59.8 % of the non treated bran, and the F4 fraction from straw digestibility increased from 44.9 to 56.7 %.

As shown in Table 6.1, the fraction F4 of wheat straw (15 % of total processed material) has a high concentration of nutrients: Consequently, an increase of nutritional value measured in UFL of 25 %. This was more evident in corn straw, where the increment was of 33 % (Table 6.2; Dell'Omo et al. 2011). These data are extremely interesting both for Developing Countries, where these by-products are the only feed available, and for the developed countries to reduce production costs and to optimize the resources.

6.3.3 Human Nutrition

Today, the by-products of milling industries (bran, wheat flour middlings) are exclusively destined for animal feed. However, the HYST treatment of such by-prod-

Table 6.3 Characteristics of HYST flour produced from bran. (Department of Animal Sciences, Milan University, 2011)

Chemical composition (% d.m.)		
		According to Regulation (EC) No. 1924/2006 (EU 2006)
Protein	21–24%	High protein content
Starch	48–55%	
Lipids	3.3–3.5%	
Fibers	4.8–15%	
Minerals	3.5–4%	
Vitamin and mineral content (% d.m.)		
Vitamin E	1.39 mg/100 g	Source of vitamin E
Thiamin (Vit. B1)	0.93 mg/100 g	Source of thiamin
Niacin (Vit. B3)	18.3 mg/100 g	Source of niacin
Pantothenic Acid (B5)	2.1 mg/100 g	Source of pantothenic acid
Folic acid (Vit. B9)	77 μg/100 g	Source of folic acid
Iron	9.2 mg/100 g	Source of iron
Zinc	3.08 mg/100 g	Source of zinc
Magnesium	271 mg/100 g	Source of magnesium

ucts produces protein flours meeting high nutritional requirements and thus eligible to be used in deficiency situation caused by malnutrition and/or undernutrition.

The innovative features of HYST technology allow to optimize the use of raw materials, creating new natural food with a high nutritional profile without introducing any toxic or chemical compound. Processing with the HYST system by-products of the milling industry, now used only for the production of feed, a flour can be obtained in which vitamins and minerals are more abundant than in the common cereal flours with a product yield of 15–20 %.

The analytical results show that the flour obtained from soft wheat bran by the HYST technology has an high protein content (up to 24 %), when the conventional wheat flour has a maximum average content of 14 %. According to Regulation CE 1924/06 (EU 2006) HYST flour can be labeled as having high protein content and source of all the micronutrients. In fact, 100 grams of HYST flour provide the recommended daily dose of vitamin B3, which is essential for metabolic functions, and over 60 % of the daily dose of vitamin B1 (Table 6.3), which is recognized by the European Food Safety Authority as a promoter of brain function in children.

These results are particularly interesting if compared to the artificially fortified flours, as for example a flour added according to USDA standards, as shown in Fig. 6.4. With its exceptional vitamins and minerals content and it completely natural origin, the HYST flour can be normally consumed as common food, but can also fit the functional food market requirements, and, most importantly, open new perspectives in solving the food scarcity problem.

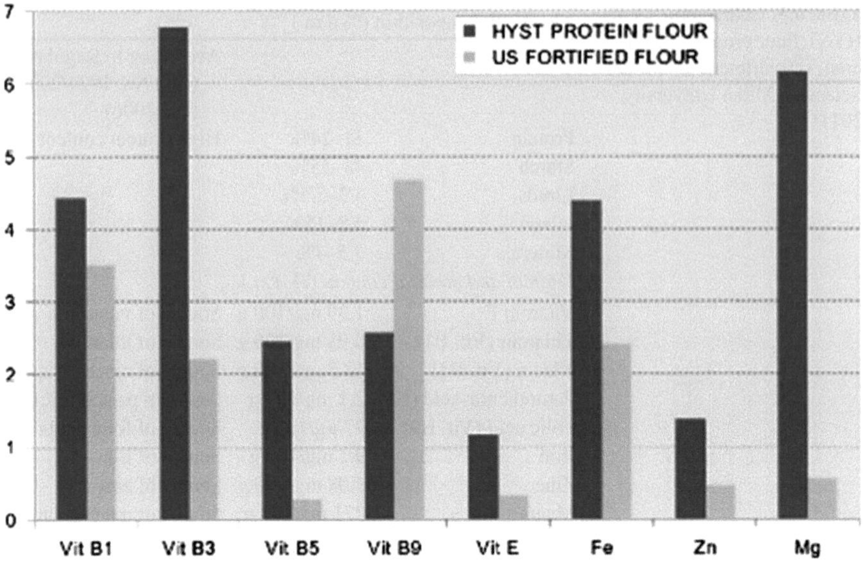

Fig. 6.4 Micronutrient content of HYST flour: comparison with artificially fortified foods (values parameterized to the significant content according to EU legislation)

6.4 Future Perspectives

Future research goals will be to optimize the HYST process on the studied agricultural by-products and applications, but also to explore the potentialities of other interesting residues.

6.4.1 Human Nutrition

The preliminary studies for the nutritional and functional characterization have been carried out on products obtained from the processing HYST bran from wheat residues. Future researches will be focused on other agricultural by-products of interest, considering also the possible experimentation on specific population groups.

Wheat semolina is a typical product of Italian agriculture, and its bran could be the first analyzed for the easy availability and large market.

Rice husks contains high amounts of beneficial antioxidants including tocopherols, tocotrienols, and oryzanols. Current rice milling technology produces rice bran from different layers of the kernel caryopsis. Under current practices, these layers are combined and then steam-extruded to form a stabilized rice husks pellet that is storage-safe prior to oil extraction (Lloyd et al. 2000). Recently, FAO reported that the annual production of rice husk in the world amounts to 4.3 million tonnes.

The main components of the husks are oils, proteins and carbohydrates. It is then used for the extraction of oils, and as an ingredient in food for humans and feed for animals. The extraction of oils from rice husks, which currently is conducted almost exclusively in Asia and in the United States, involves the use of n-hexane, which can be emitted into the atmosphere and has a negative impact on the environment. The extracted oil, subsequently, before being used requires further extraction steps. Other enzymatic processes provide the simultaneous extraction of proteins and fats from rice husk (Hanmoungjai et al. 2002).

Furthermore, proteins extracted from rice, are considered extremely interesting as functional ingredients (Tang et al. 2003). Thanks to the high content of lysine, the protein derived from rice husk are hypoallergenic (Helm and Burks 1996), and this can be used as an ingredient in food formulations for children with allergies (Burks and Helm 1994). The amino acid profile of rice proteins can be considered better than that of casein and soy as it corresponds to amino acid needs of children aged 2–5 years (Wang et al. 1999).

The main goal of the compositional study of the HYST processed rice husk products could be the development of a flour enriched in amino acids and antioxidants to be used alone or mixed with other traditional flours for specific population groups (e.g. malnourished children for its amino acid profile).

There is also an increasing interest in the exploitation of residues generated by the wine industry (Arvanitoyannis et al. 2006). In particular, the grape pomace may constitute an alternative for the extraction of natural antioxidant compounds, which are considered to be entirely safe compared to synthetic antioxidants. The skins are a rich source of molecules of great interest, including oil, hydrocolloids and dietary fiber. Furthermore, the grape pomace is characterized by a high content of polyphenolic compounds, which are not extracted during the process of wine production (Kammerer et al. 2004). The waste products obtained after the production of wine, constitute a very economical source for the extraction of matrices with a high content of antioxidants (Alonso et al. 2002; Negro et al. 2003; Gonzalez-Paramas et al. 2004). The identification of cheap technologies for the processing of skins could be a significant boost to their exploitation.

The application of the HYST process on this kind of residues would have the aim of obtaining fine fractions high in dietary fiber and polyphenolic compounds, to be used as such or as functional ingredients to be added to flour or other food, after the selection of the more suitable grape pomace (in terms of cultivar and kind of production).

6.4.2 *Green Chemistry*

The chemical industry is showing a growing interest in the possibility of obtaining chemicals from renewable raw materials (biochemicals) in alternative to the existing products, derived mostly from fossil fuels. The "green chemistry" has many advantages: almost inexhaustible availability of renewable resources, lower environmental impact, biodegradability of products, reduction of CO_2 emissions, lower energy consumption and lower raw material.

In particular the conversion of lignocellulosic biomass into oligo-and monosaccharides for the production of fuels and chemicals by microorganisms, is an extremely attractive option for the petrochemical industry for the production of a new generation of chemical molecules base. Researchers from the U.S. Department of Energy (DOE) have identified numerous building blocks which may be obtainable from biomass (Werpy et al. 2004).

Among the building blocks of renewable origin listed by US DOE there are several organic acids (succinic, maleic, aspartic, furanic, itaconic, levulinic, glucanic) that can be used for the production of bioplastics, biolubricants and biosolvents.

HYST process could be applied in the pretreatment step for the production of succinic and lactic acids from lignocellulosic biomass to facilitate the difficult hydrolysis phase. The results will be economically evaluated in order to compare the production costs to those produced by fermentative process of cereals, as the succinic acid or of petrochemical origin as lactic acid.

References

Alonso A, Guillean D, Barroso C, Puertas B, Garcia A (2002) Determination of antioxidant activity of wine by-products and its correlation with polyphenolic content. J Agr Food Chem 50:5832–5836

Arvanitoyannis IS, Ladas D, Mavromatis A (2006) Potential uses and applications of treated wine waste: a review. Int J Food Sci Tech 41:475–487

Burks AW, Helm R (1994) Hypoallergenicity of rice protein. Presented at the Annual Meeting of the American Association of Cereal Chemists, Nashville, TN

Dell'Omo P, Luciani F, Malagutti L (2011) HYST treatment of biomass: results for animal feeding and bioenergy. Tecnica Molitoria 62(7):708–715

Department of Animal Sciences, Milan University (2011) Department of Animal Sciences, Milan University—Analysis Report dated April 21, 2011. www.biohyst.it. Accessed 25 Oct 2012

ENEA (2006) Dossier ENEA: Le tecnologie per i biocombustibili e i biocarburanti: opportunità e prospettive per l'Italia. A cura di Vito Pignatelli. http://www.sede.enea.it/produzione_scientifica/pdf_dossier/D04_Biocombustibili.pdf. Accessed 15 Oct 2012

EU (2006) Regulation (EC) No 1924/2006 of the European Parliament and of the Council of 20 December 2006 on nutrition and health claims made on foods. Official Journal of the European Union 30.12.2006

EU (2008) Directive 2008/98/EC of the European Parliament and of the Council of 19 November 2008 on waste and repealing certain Directives. Official Journal of the European Union 22.11.2008

EU (2009) Directive 2009/28/EC of the European Parliament and of the Council of 23 April 2009 on the promotion of the use of energy from renewable sources and amending and subsequently repealing Directives 2001/77/EC and 2003/30/EC. Official Journal of the European Union L 140/16

Gonzalez-Paramas A, Esteban-Ruano S, Santos-Buelga C, Pascual-Teresa S, Rivas-Gonzalo J (2004) Flavanol content and antioxidant activity in winery by-products. J Agr Food Chem 52:234–238

Hanmoungjai PJ, Pyle DL, Niranjan K (2002) Enzyme-assisted water-extraction of oil and protein from rice bran. Chem Technol Biotechnol 77:771–776

Helm RM, Burks AW (1996) Hypoallergenicity of rice protein. Cereal Food World 41:839–843

Kamalak A, Filbo JMP, Canbolat O, Gurbuz Y, Ozay O, Ozkan CO (2004) Chemical composition and its relationship to in vitro dry matter digestibility of several tannin-containing trees and shrub leaves. Livest Res Rural Dev 16(6). http://www.lrrd.cipav.org.co/lrrd16/6/kama16044.htm. Accessed 4 Nov 2012

Kammerer D, Claus A, Carle R, Schieber A (2004) Polyphenol screening of pomace from red and white grape varieties (Vitis vinifera L.) by HPLC-DAD-MS/MS. J Agr Food Chem 52:4360–4367

Lloyd BJ, Siebenmorgen TJ, Beers KW (2000) Effects of commercial processing on antioxidants in rice bran. Cereal Chem J 77:551–555

Menke KH, Steingass H (1988) Estimation of the energetic feed value obtained from chemical analysis and in vitro gas production using rumen fluid. Anim Res Dev 28:7–55

Negro C, Tommasi L, Miceli A (2003) Phenolic compounds and antioxidant activity from red grape marc extracts. Bioresource Technol 87:41–44

Ruberto G, Renda A, Daquino C, Amico V, Spatafora C, Tringali C, De Tommasi N (2007) Polyphenol constituents and antioxidant activity of grape pomace extracts from five Sicilian red grape cultivars. Food Chem 100:203–210

Tang S, Hettiarachchy NS, Horax R, Eswaranandam S (2003) Physicochemical properties and functionality of rice bran protein hydrolyzate prepared from heat-stabilized defatted rice bran with the aid of enzymes. J Food Sci 68:152–157

Unione P (2012) Previsioni di domanda energetica e petrolifera italiana 2012–2025. http://www.unionepetrolifera.it/it/CMS/pubblicazioni/get/2012/Previsioni%20UP%202012_2025.pdf. Accessed 3 Nov 2012

Wang M, Hettiarachchy NS, Qi M, Burks W, Siebenmorgen T (1999) Preparation and functional properties of rice bran protein isolate. J Agr Food Chem 47:411–416

Werpy T et al (2004) Top value added chemicals from biomass, Volume 1—Results of screening for potential candidates from sugars and synthesis gas 2004, report NREL/TP-510-355532. US Department of Energy

Chapter 7
Antioxidative Effectiveness of Environment Friendly Functional Biopolymers for Food Applications

U. Gianfranco Spizzirri, Donatella Restuccia, Giuseppe Cirillo, Francesco Puoci, Ortensia I. Parisi and Nevio Picci

Abstract As consumers are demanding polymeric materials more natural, disposable, biodegradable and recyclable, the interest in active food-packaging systems in which the polymer matrix has low environmental impact is growing significantly. To this regard, the insertion of antioxidant molecules onto polysaccharides backbone is an innovative strategy to produce biocompatible species suitable for food-packaging applications. Moreover, free radical grafting is a method which allows the synthesis of antioxidant biomacromolecule without the use of organic solvents or toxic radical initiators. In the present study, quercetin, catechin and gallic acid were bound to chitosan, starch, inulin and alginate using this technique and good antioxidant activity of the conjugates were observed. This can be useful in the optimization of food preservation and to help manufacturers in elaboration of new and sustainable food products and packaging.

Keywords Functional biopolymers · Environment friendly · Antioxidant · Free radical grafting · Redox initiator

7.1 Introduction

The use of proper packaging materials and methods to minimize food losses and provide safe and wholesome food products, has always been the focus of food packaging. In addition, consumer trends for better quality, fresh-like, and convenient food products have intensified during the last decade. Therefore, a variety of active packaging technologies have been developed to provide better quality, safe foods and also to limit package related environmental pollution and disposal problems. In this sense, biodegradable active packaging represent novel food packaging systems, as they can play a proactive role regarding product preservation, shelf-life extension, and even improvement.

Among active packaging classes, polymeric antioxidants are macromolecular systems characterized by higher stability and slower degradation rate than compounds

D. Restuccia (✉) · U. G. Spizzirri · G. Cirillo · F. Puoci · O. I. Parisi · N. Picci
Department of Pharmacy, Health and Nutritional Sciences, University of Calabria, 87036 Rende, Italy
e-mail: donatella.restuccia@unical.it

R. Salomone, G. Saija (eds.), *Pathways to Environmental Sustainability*,
DOI 10.1007/978-3-319-03826-1_7,

with low molecular weight. The possibility to graft antioxidant moieties in a macromolecular structure, by radical procedure, represents an interesting innovation which significantly improves the performance of the biomacromolecules, opening new applications in the food industries as new functional food and/or innovative materials for food packaging applications. Therefore, the conjugation between antioxidant molecules and biopolymer combines the advantages of both the components: the higher stability and slower degradation rate of molecules with high molecular weight and the unique properties of antioxidant molecules (Puoci et al. 2008a).

Basically, the traditional procedures to synthesize antioxidant–polymer conjugates consist to functionalize the biomacromolecules by drastic chemical conditions (high temperature, extreme pH conditions, etc.). Furthermore, these processes were often performed in organic solvent, reducing the potential use of the modified materials. To face these problems, is possible to functionalize the polymeric materials under mild and eco-friendly conditions exploiting the advantages of the grafting procedure. Graft polymerization is a well-known method for the modification of chemical and physical structure to tailor properties for specific application. This method is of particular interest to achieve specifically designed polymer properties by connecting different types of polymers having the desired properties in the same polymer chain (Nurmi et al. 2007). The advantages in the use of the grafting reaction consist in the functionalization of the materials without affecting its bulk properties, elimination of the need to redesign the bulk material to achieve a target surface performance and a considerably reduction of the cost.

Moreover, this synthetic strategy can be applied to natural macromolecules to meet the growing public concern about environmental pollution leading to development and design of biodegradable polymeric materials. In particular, natural polysaccharides, such as chitosan, starch, alginate and inulin are widely employed in industry due to their biocompatibility, biodegradation, non-toxicity and non-immunogenicity and grafting can very useful to synthesize polysaccharide–antioxidant conjugates using renewable raw materials, without the generation of toxic reaction by-products and at room temperatures, preserving the antioxidant by degradation processes (Cirillo et al. 2010). The key advantage of these techniques is that the surface of the materials can be modified or tailored to acquire very distinctive properties through the choice of different grafting bioactive molecules, while maintaining the substrate properties. Compared with the physically coated polymer chains, the covalent attachment of the grafted chains onto a material surface avoids their desorption and maintains a long-term chemical stability of the introduced chains. Surface grafting commonly includes two steps: surface activation and graft reaction. Because of the absence of chemically reactive functional groups on most substrate surfaces, a surface activation process is indeed needed to create reactive sites on them that can generate further grafting processes.

The interest in these kind of systems is confirmed by the Economy Watch stating that the market for biodegradable packaging has been developing rapidly over the last decade with estimations suggesting an annual growth of greater than 20%. In the next several years, the market demand for biodegradable packaging materials will continue to grow, reaching $ 20 billion by 2020 with an average annual growth rate of 22 %. Global bioplastic packaging demand is forecast to reach 884,000 t by 2020. A 24.9 % CAGR is expected from 2010 to 2015 slowing to 18.3 % in the 5 years to 2020. Europe is the largest regional market for bioplastic packaging with over half of world tonnage in 2010. It benefits from favorable consumer and retail attitudes to sustainable packaging, supportive government policies towards packaging waste recycling and a well-developed composting infrastructure.

In this chapter the grafting of antioxidant molecules (catechin, quercetin and gallic acid) onto the polysaccharidic structures above mentioned allowed to obtain new functionalized materials characterized by the properties of both grafted molecule and natural polymer. In the discussion, the covalent insertion of these antioxidant molecules onto macromolecular systems by employing free radical grafting reaction in which a water soluble redox pair as initiator system was deeply investigated. The proposed approach can be useful in the optimization of food preservation and to help manufacturers in elaboration of new and sustainable active packaging systems.

7.2 Natural Polymers

7.2.1 Chitosan

Chitosan is a copolymer of N-acetyl-D-glucosamine and D-glucosamine obtained by alkaline N-deacetylation of chitin. The sugar backbone consists of β-1,4-linked glucosamine, and it has been known as a bioactive molecule (Chae et al. 2005). Several bioactivities such as antitumor activity (Pae et al. 2001), immunoenhancing effects (Maeda et al. 1992), wound healing effects (Porporatto et al. 2003), antifungal and antimicrobial properties (Hirano 1996), and antioxidant activity (Yen et al. 2008) of chitosan have been reported. These characteristics, together with several unique properties such as nontoxicity, biocompatibility, and biodegradability, offer chitosan good potential for biomedical applications, in the food industry as an edible coating for fruits and vegetables (Park 1999) or packaging film (Caner et al. 1998), and in wastewater purification (Knorr 1991).

7.2.2 Inulin

Inulin is a dietary fiber composed of a mixture of oligo- and/or polysaccharides consist of fructose unit chains (linked by (2→1)-β-D-fructosyl-fructose bonds) of various length, terminated generally by a single glucose unit (linked by an α-D-glucopyranosyl bond) (Roberfroid and Delzenne 1998). Inulin is a prebiotic, a selectively fermented ingredient that allows specific changes, both in the composition and/or activity in the gastrointestinal microbiota, that confers benefits upon host well-being and health (Gibson et al. 2004). The use of inulin in food formulations as fiber ingredient to influence the composition of colonic microflora is straightforward and often leads to significantly improved organoleptic characteristics, such as taste, mouthfeel, and texture. The insertion of a biocompatible antioxidant agent onto the structure of a prebiotic could be interesting to improve the stability of this kind of food ingredients.

7.2.3 Starch

Starch or amylum is a carbohydrate consisting of a large number of glucose units joined by glycosidic bonds. This polysaccharide is produced by all green plants as an energy store. It is the most common carbohydrate in the human diet and is contained in large amounts in such staple foods as potatoes, wheat, maize (corn), rice, and cassava. One area among many that could be of particular interest is the use of starch as a renewable raw material for chemical conversion to degradable plastic materials (Morell and Myers 2005). In addition, starch has been investigated widely for the potential manufacture of products such as water soluble pouches for detergents and insecticides, flushable liners and bags, medical delivery systems and devices (Ma et al. 2008) like fillers, binders and disintegrants in the pharmaceutical fields because it is cost-effective and non-toxic (Gibson et al. 2004).

7.2.4 Alginate

Alginates represent a class of linear unbranched polysaccharides which contain varying amounts of 1,4′-linked β-D-mannuronic acid and α-L-guluronic acid residues. The residues may vary widely in composition and sequence and are arranged in a pattern of blocks along the chain. Alginates and their derivatives are widely used in food and beverage industries, as thickening agents, gelling agents and colloidal stabilizers (Babu et al. 2007) and agriculture applications (Puoci et al. 2008b). In literature, many studies report on the use of alginate coatings in packaging technology to improve quality and shelf life of food (Lu et al. 2009). Alginate is a renewable, biocompatible (Klöck et al. 1997) and biodegradable natural polymer that is used in a variety of commercial applications because of its capacity to form hydrogels.

7.3 Grafting Procedure

Literature data report on the introduction of particular functionalities in a macromolecular chain grafting by chemical reaction involving as initiators potassium persulfate, ceric ammonium nitrate and ammonium persulfate, producing free radical species after warming at 40 °C (Mishra et al. 2008). In this work, novel polysaccharide with antioxidant properties were synthesized by a simple method, involving a one-step reaction, to obtain new biomacromolecules through the simple conjugation

Table 7.1 Chemical composition of antioxidant–polysaccharide conjugates

Polysaccharides	Antioxidant	Solvent	Initiator system
Chitosan (0.5 g)	Gallic acid (0.35 mmol) Catechin (0.35 mmol)	10 ml of acetic acid water solution (2 % v/v)	1 ml of 1.0M H_2O_2 and 0.054 g of ascorbic acid
Inulin (1.5 g)	Catechin (0.35 mmol)	37.5 ml distilled water	12.5 ml H_2O_2 (120 v) and 0.4 g of ascorbic acid
Alginate (1.5 g)	Catechin (0.35 mmol)	37.5 ml distilled water	12.5 ml H_2O_2 (120 v) and 0.4 g of ascorbic acid
Starch (1.0 g)	Quercetin (0.16 mmol)	13 ml water/ethanol (5/5, v/v) mixture	1.5 ml of H_2O_2 1.0 M and 0.081 g of ascorbic acid

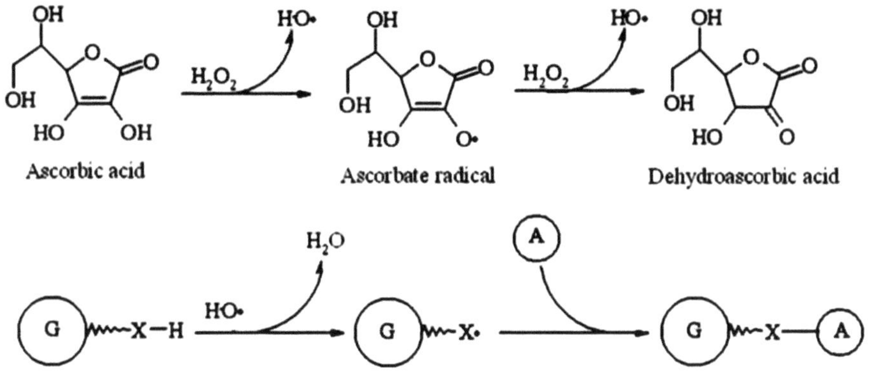

Fig. 7.1 Interaction between ascorbic acid and hydrogen peroxide and insertion of antioxidant onto polysaccharide backbones

of polymer chain with an antioxidant molecule in the presence of water-soluble redox initiators able to generate free radical species at room temperature. The addition of antioxidant molecules on the polysaccharide side chains took place by radical reaction of the biomacromolecule and antioxidants, and using the hydrogen peroxide/ascorbic acid redox pair, as initiator system (Spizzirri et al. 2009; Simms and Cunningham 2007). In particular, different antioxidant biomacromolecules were synthesized employing chitosan, inulin, alginate and starch as polysaccharides and gallic acid, catechin and quercetin as antioxidant molecules (Table 7.1). Blank polymers, acting as controls, were prepared in the same conditions but in the absence of the antioxidant molecules.

A possible mechanism to bind antioxidant molecules to polysaccharide side chains is proposed in Fig. 7.1.

The hydroxyl radicals, generated by the interaction between redox pair components, attack the sensible residues in the side chains of the polysaccharides, producing radical species on the sugar structures. These ones react with the antioxidant molecules inducing an antioxidant–polysaccharide covalent bond. Literature data suggests that free radical species attack at the ortho- and para-positions relatively to the hydroxyl group on the phenolic ring (Kobayashi and Higashimura 2003).

The hydroxyl radicals, generated by the interaction between redox pair components, attack the sensible residues in the side chains of the polysaccharides, producing radical species on the sugar structures. These ones react with the antioxidant molecules inducing an antioxidant–polysaccharide covalent bond. Literature data suggests that free radical species attack at the ortho- and para-positions relatively to the hydroxyl group on the phenolic ring (Kobayashi and Higashimura 2003).

7.4 Polysaccharide–Antioxidant Conjugates Characterization

The covalent insertion of antioxidant in the polymeric chain was confirmed by infrared, calorimetric and fluorescence analyses, while a crude estimation of the amount of antioxidant bound per gram of polymer was obtained by performing the Folin–Ciocalteu assay. The antioxidant activity of conjugated polymers was tested and functionalized macromolecules were able to interact with free radical species and to minimize the oxidative damage.

7.4.1 Physico-Chemical Properties

IR-spectra of all antioxidant conjugates showed the appearance of new peaks awardable to the functionalities of antioxidant moieties. A further confirmation of antioxidants insertion in the biopolymer was obtained by comparing UV and emission spectra of each antioxidant molecule and the respective conjugates in water. The bathochromic shifts of the aromatic peaks were used as prove of the covalent linkage of the antioxidants with the polysaccharide. Finally, calorimetric analyses of pure antioxidants, blank polymers and conjugates were performed to check the conjugates to be free by un-reacted antioxidants.

7.4.2 Antioxidant Properties

A more specific characterization of the synthesized conjugates was performed to evaluate their antioxidant properties by three specific antioxidant tests and the experimental data are reported in Table 7.2. The determination of total antioxidant activity is based on the reduction of Mo(VI) to Mo(V) by ferulic acid and subsequent formation of a green phosphate/Mo(V) complex at acid pH (Prieto et al. 1999). The total antioxidant activity was measured and compared with that of antioxidants and the respective blank polymers, which did not contain antioxidant component.

Table 7.2 Disposable phenolic groups and antioxidant activity of polysaccharides–antioxidant conjugates

Antioxidant conjugate	Antioxidant activity			
	TAC*	DPG*	Scavenger effect against DPPH radical inhibition (%)	β-Carotene-Linoleic acid assay inhibition (%)
Chitosan–gallic acid	18.2±0.4	31.1±0.5	92.0±1.2	85.2±0.9
Chitosan–catechin	17.2±0.2	23.5±0.3	95.3±1.0	95.0±1.0
Inulin–catechin	4.9±0.2	2.7±0.2	74.0±1.2	60.0±0.8
Alginate–catechin	1.9±0.1	1.1±0.1	61.0±0.9	50.0±0.9
Starch–quercetin	32.0±0.8	41.6±1.2	99.0±1.0	99.1±1.0

TAC total antioxidant activity, *DPG* disposable phenolic groups, *DPPH* diphenylpicrylhydrazine
* Expressed as μmol antioxidant per gram of polymer

The experimental data, reported in Table 7.2, indicated that the samples possessed significant antioxidant activity, which was expressed as μmol equivalent of antioxidant per gram of dry conjugate. In particular, the starch–quercetin conjugate showed the highest total antioxidant capacity while the conjugation of catechin with the alginate produced a macromolecular system 16.8-fold less efficient against oxidant species.

The DPPH radical is a stable organic free radical with an absorption maximum band around 515–528 nm and thus, it is a useful reagent for evaluation of antioxidant properties of compounds. In the DPPH assay, the antioxidants reduce the DPPH radical to a yellow-colored compound, diphenylpicrylhydrazine, and the extent of the reaction will depend on the hydrogen donating ability of the antioxidants. Conjugates scavenger ability were evaluated in term of DPPH reduction and data are expressed as inhibition (%), as reported in Table 7.2, and, in our operating conditions, all antioxidant polymers showed high scavenging activity with starch and chitosan conjugates totally inhibiting the DPPH radical, while partial inhibition was observed for alginate and inulin.

A further characterization of the antioxidant properties was performed in terms of lipid peroxidation inhibition. Linoleic acid was used as a substrate and *β*-carotene as detecting agent. In the *β*-carotene–linoleate model system, one hydrogen atom of linoleic acid is withdrawn, leaving a free radical ready to attack *β*-carotene molecules. In this reaction, *β*-carotene molecules lose the double bond and their characteristic orange color fades, and this oxidative degradation of *β*-carotenes by radicals on linoleic acid is measured by the decrease in absorbance at 470 nm (Kodali and Sen 2008). The high antioxidant properties of the conjugates, reported in Table 7.2, as inhibition (%) is clearly understandable. The data confirmed the trend observed against the DPPH radical and, also in this case, the each respective blank polysaccharides were ineffective.

Since the antioxidant activity of conjugates is derived from phenolic groups in the polymeric backbone, it is useful express the antioxidant potential in terms of phenolic content. The Folin–Ciocalteu phenol reagent is used to obtain a crude estimate of the amount of disposable phenolic groups present in polymer chain. Phenolic compounds undergo a complex redox reaction with phosphotungstic and phosphomolybdic acids present in the Folin–Ciocalteu reactant. The color development is due to the transfer of electrons at basic pH to reduce the phosphomolybdic/phosphotungstic acid complexes to form chromogens in which the metals have lower valence (Spizzirri et al. 2009). For each grafted polymer, disposable phenolic groups were expressed as mg equivalent of polyphenol by comparing the obtained data with the calibration curve recorded for each antioxidant (Table 7.2). The experimental data showed different amount of phenolic groups in the polymeric chains confirming the same trend observed for the antioxidant activity.

In addition, the conjugate starch–quercetin, showing the best antioxidant properties, was investigated to evaluate the effect of the covalent insertion of the antioxidant in the polymeric backbone on the polymer stability. To this regard, a degradation test induced by UV treatment on both conjugated and blank starch was carried out. When polymer materials are irradiated by UV light, they are usually destroyed easily. The small molecules produced after UV-irradiation, such as hydrogen gas, carbon dioxide, carbon monoxide, water, etc., will evaporate from the material matrix; as a consequence, the weight of the materials will become smaller. Our hypothesis was that the covalent conjugation of starch with the flavonoid should improve the UV stability of the starch, and the obtained results confirmed the enhanced stability of the synthesized macromolecular system, with a degradation value (%) which decreases from $24 \pm 1.3\,\%$ for blank starch to $9 \pm 1.5\,\%$ for the conjugate.

7.5 Conclusion

There is a worldwide realization of the damage that a rapidly rising indiscriminate industrial processes can do to the ecology and environmental balance of the planet. There is general agreement that future technology development will need concepts such as biological sustainability, minimum use of energy and renewable raw materials that will probably be set internationally. Current R&D needs emphasizes on the development of high-value and safer products. In this sense, polymeric antioxidants based on natural polymers are a very promising class of materials with a great potential in many research and industrial areas. The high stability, the biocompatibility and the good biological properties make these systems suitable for several food applications, such as preservatives, food packaging, etc.

Green chemistry concepts may be utilized for eco-friendly materials and to minimize the toxic implications of organics which pose potential risks to human health. To this regard, free radical grafting is a straightforward synthetic procedure to obtain polymeric antioxidants with high yields and avoiding the use of organic solvents and the generation of toxic by-products.

References

Babu VR, Sairam M, Hosamani KM, Aminabhavi TM (2007) Preparation of sodium alginate–methylcellulose blend microspheres for controlled release of nifedipine. Carbohyd Polym 69:241–250

Caner C, Vergano PJ, Wiles JL (1998) Chitosan film mechanical permeation properties as affected by acid, plasticizer and storage. J Food Sci 63:1049–1053

Chae SY, Jang M, Nah J (2005) Influence of molecular weight on oral absorption of water soluble chitosans. J Control Release 102:383–394

Cirillo G, Kraemer K, Fuessel S, Puoci F, Curcio M, Spizirri UG, Altimari I, Iemma F (2010) Biological acitvity of gallic acid-gelatin conjugate. Biomacromolecules 11:3909–3915

Gibson GR, Probert HM, Van Loo J, Rastall RA, Roberfroid MB (2004) Dietary modulation of the human colonic microbiota: updating the concept of prebiotics. Nutr Res Rev 17:259–275

Hirano S (1996) Chitin biotechnology applications. Biotechnol Annu Rev 2:237–258

Knorr D (1991) Recovery and utilization of chitin and chitosan in food processing waste management. Food Technol 45:114–122

Kobayashi S, Higashimura H (2003) Oxidative polymerization of phenols revisited. Prog Polym Sci 28:1015–1021

Kodali VP, Sen R (2008) Antioxidant and free radical scavenging activities of an exopolysaccharide from a probiotic bacterium. Biotechnol J 3:245–251

Lu F, Liu D, Ye X, Wei Y, Liu F (2009) Alginate–calcium coating incorporating nisin and EDTA maintains the quality of fresh northern snakehead (*Channa argus*) fillets stored at 4 °C. J Sci Food Agr 89:848–854

Ma X, Jian R, Chang PR, Yu J (2008) Fabrication and characterization of citric acid-modified starch nanoparticles/plasticized-starch composites. Biomacromolecules 9:3314–3320

Maeda, Murakami H, Ohta H, Tajima M (1992) Stimulation of IgM production in human-human hybridoma HB4C5 cells by chitosan. Biosci Biotechnol Biochem 56:427–431

Mishra A, Clark JHVA, Daswal S (2008) Synthesis of graft copolymers of xyloglucan and acrylonitrile. Polym Advan Technol 19:99–104

Morell MK, Myers AM (2005) Towards the rational design of cereal starches. Curr Opin Plant Bio 8:204–210

Nurmi L, Holappa S, Mikkonen H, Seppälä J (2007) Controlled grafting of acetylated starch by atom transfer radical polymerization of MMA. Eur Polym J 43:1372–1382

Pae HO, Seo WG, Kim NY, Oh GS, Kim GE, Kim YH, Kwak HJ, Yun YG, Jun CD, Chung HT (2001) Induction of granulocytic differentiation in acute promyeloocytic leukemia cells (HL 60) by water-soluble chitosan oligomer. Leukemia Res 25:339–346

Park HJ (1999) Development of advanced edible coatings for fruits. Trends Food Sci Technol 10:254–260

Porporatto C, Bianco ID, Riera CM, Correa SG (2003) Chitosan induces different l-arginine metabolic pathways in resting and inflammatory macrophages. Biochem Biophys Res Commun 304:266–272

Prieto P, Pineda M, Aguilar M (1999) Spectrophotometric quantitation of antioxidant capacity through the formation of a phosphomolybdenum complex: specific application to the determination of vitamin E. Anal Biochem 269:337–341

Puoci F, Iemma F, Curcio M, Parisi OI, Cirillo G, Spizzirri UG, Picci N (2008a) Synthesis of methacrylic-ferulic acid copolymer with antioxidant properties by single-step free radical polymerization. J Agric Food Chem 56:10646–10650

Puoci F, Iemma F, Spizzirri UG, Cirillo G, Curcio M, Picci N (2008b) Polymer in agriculture: a review. Am J Agr Biol Sci 3:299–314

Roberfroid MB, Delzenne NM (1998) Dietary fructans. Ann Rev Nutr 18:117–143

Simms RW, Cunningham MF (2007) High molecular weight poly(butyl methacrylate) by reverse atom transfer radical polymerization in miniemulsion initiated by a redox system. Macromolecules 40:860–866

Spizzirri UG, Iemma F, Puoci F, Cirillo G, Curcio M, Parisi OI, Picci N (2009) Synthesis of antioxidant polymers by grafting of gallic acid and catechin on gelatin. Biomacromolecules 10:1923–1930

Yen M, Yang J, Mau J (2008) Antioxidant properties of chitosan from crab shells. Carbohydr Polym 74:840–844

Part II
Experiences and Case Studies in Productive Contexts

Chapter 8
1997–2012 The Environmental Management System of the ST Microelectronics Site of Catania is 15 years old

Antonio Battiato

Abstract Lead by the insight of Pasquale Pistorio, the man who claimed a leading role for the electronic semiconductor industry in the European macroeconomic system, ST Microelectronics attempted to assume the role of "defender of the environment"; we can say with encouraging results. Aiming not, to debate the subject in exhaustive manner, this short text highlights some data from the ST site of Catania, showing the advantages that an environmental management system can produce when it is deployed in a manufacturing organization. Furthermore, what we most want, that the "systematic approach" will spread to the other industrial organizations but also to the institutions, to the research and training organizations, to the public and private administrators, in short, to every citizen aware that economic development is impossible without social equity and environmental protection.

Keywords Environment · Management system · EMAS · Semiconductor industry · ST Microelectronics

8.1 The ST Microelectronics—History—Products—Processes

STMicroelectronics is an independent global semiconductor maker which designs, develops, manufactures and sells a wide spectrum of electronics applications with innovative semiconductor solutions (e.g. multimedia and power applications, devices for automotive industry, etc.).

STMicroelectronics, using its huge array of technologies, design expertise and combination of intellectual property portfolio, strategic partnerships and manufacturing strength, holds today the seventh position in the worldwide revenue ranking for semiconductor manufacturers (Company's net revenue in 2011 was US$ 9,735 million).

SGS-Thomson was created in June 1987 by the merger of two long—established semiconductor companies; the Italian SGS Microelettronica (Società Generale Semiconduttori) and the semiconductor arm of the French company, Thomson. The

A. Battiato (✉)
ST Microelectronics—site of Catania, 95121 Catania, Italy
e-mail: antonio.battiato@st.com

R. Salomone, G. Saija (eds.), *Pathways to Environmental Sustainability,*
DOI 10.1007/978-3-319-03826-1_8, © Springer International Publishing Switzerland 2014

new European corporation took the actual name of STMicroelectronics (shortly ST) in 1998.

Since it has been publicly traded in 1994 (Stock Exchange of New York, Paris and Milan) STMicroelectronics has enlarged its products and technologies portfolio and enhanced its production capacities improving subsequently its place in the ranking of global revenue for semiconductor manufacturers up to the second place just after Intel in 2001. The group has about 50,000 employees, 12 manufacturing plants, 7 main R&D technology centers, 39 design centers and sales offices around the world.

Semiconductor device fabrication is the process used to create the integrated circuits that are present in everyday electrical and electronic devices. It is a multiple-step sequence of photolithographic and chemical processing phases during which electronic circuits are gradually created on a disc made of pure silicon (called wafer); silicon is the most used but also other semiconductor materials are used for specialized applications.

The entire manufacturing process, from start to packaged chips ready for the market, takes 6–8 weeks and is performed in highly specialized facilities.

Once the wafers are prepared, the various process steps necessary to produce the desired semiconductor integrated circuit can be divided into two major groups: front-end and back-end processing. ST factories are devoted to the first or the second of these process groups.

To provide its customers with an independent, secure and cost-effective manufacturing machine, ST operates a worldwide network of front-end (wafer fabrication) and back-end (assembly, test and packaging) plants. ST's main wafer fabs are presently located in Agrate Brianza and Catania (Italy), Crolles, Rousset and Tours (France), and in Singapore, while the back-end facilities are located in China, Malaysia, Malta, Morocco, the Philippines and Singapore.

The devices produced by the company are designed and manufactured with many different processes and design methodologies created internally. The Company's world-class products and technologies serve to:

- enable the production of multimedia and communication smart devices that help people to interact anywhere, anytime;
- increase energy efficiency all along the energy chain, from power generation to distribution and consumption;
- provide all aspects of data security and protection;
- contribute to help people live longer and better by enabling emerging healthcare and wellness applications.

The front-end processing is the first portion of IC (integrated circuits) fabrication, where the individual devices are patterned on the semiconductor surface, and it is performed in high tech facilities called clean rooms. Production of integrated circuits with very small geometries (0.1~0.2 μm) requires a very low particle contamination, and furthermore a highly controlled temperature and humidity level. Front-end processing steps fall into four general categories: deposition, removal, patterning and modification of electrical properties.

- Deposition is any process that grows, covers or otherwise transfers a material onto the wafer.
- Removal processes are those which remove material from the wafer either selectively or entirely and consist primarily of etch processes; chemical-mechanical removal processes are also used.
- Patterning refers to the series of processes that shape the deposited materials; it is generally known as lithography. For example, in conventional photolithography, the wafer is coated with a chemical called photoresist. The photoresist is exposed to short wavelength light through a mask to transfer the desired geometries on to the wafer surface.
- Modification of electrical properties has historically consisted of doping by diffusion furnaces and later by ion implantation. These doping processes are followed by furnace anneal or in advanced devices, by rapid thermal anneal which serve to activate the implanted dopants.
- Modern chips have up to 300 sequenced processing steps.
- After front-end processing there is back-end processing; it includes wafer test, wafer back grinding, die separation, die tests, IC packaging and final test. Back-end processing is done not in a cleanroom.
- The operation of a cleanroom requires large quantities of energy (for the air filtering and conditioning) large quantities of ultrapure water (wafers washing along the several processing steps) and quite large amounts of chemicals liquid and gaseous.

8.2 ST and the Environment

Under the pressure of the CEO (Chief Executive Officer) Pasquale Pistorio, in 1993 ST began to talk about ecology and environmental protection.

Graduated in electrical engineering, and specialization in electronics from the Polytechnic of Turin, Mr. Pistorio began his professional career as a semiconductor vendor just in Turin. Then he moved to Milan as director of sales for Motorola's products in a distribution company. In 1967 it became part of the commercial organization of Motorola itself and, in June 1970, he was appointed marketing director for Europe. From this time he had a fast-growing career in the American semiconductor company, becoming director of the world marketing in Phoenix (Arizona) and Vice President of Motorola Corporation in July 1977.

Shortly thereafter, in November 1978, he was appointed general manager of Motorola's semiconductor division and responsible for the design, manufacturing and marketing for all areas outside the USA.

In July 1980 his career touched a turning point when he decided to return to Italy, accepting the appointment of CEO of the SGS Microelettronica, the only Italian microelectronic company.

In May 1987 he was appointed CEO of the emerging European semiconductor group that he would lead for almost twenty years, until March 2005.

During the entire its chairmanship, Pistorio has strongly supported the cause of the European microelectronic, its possibilities and its vital importance within each macro-economic system. His greatest success, in fact, the effective integration of SGS Microelettronica and Thomson Semiconducteurs, that led ST to very high levels of technological prestige for both its broad and diversified product portfolio and for its global expansion.

The second important achievement for Pasquale Pistorio, was its tireless sponsorship to the environmental care. Among the first, he perceived that sustainable development was not just a theoretical concept but could be put in practice and translated into objectives and action plans.

As he liked to remember, he began to think about these ideas urged by the continuous requests coming from his young son, at that time university student.

At the head of a large manufacturing company, in full growth, operating in the high-tech sector, he began to note many similarities between economy and ecology and realized that companies such ST had to support the sustainable development movement. Thus he began to develop a strategy whose base was not just an ethical and social motivation, but also the firm belief that environmentally oriented industries were financially more competitive, and finally the practical need to attract and retain the best young talents.

Under his boost, in 1993 ST began to speak about ecology and environmental protection. Pasquale Pistorio started to spread his strategy throughout the organization, exploiting the fertile substrate that had been laid with TQM.

The TQM discipline, strongly focused on individual development that transforms employees from factors to actors of the production process and puts the human being at the centre of economic development, enabled the fast integration of the environmental protection into ST's corporate philosophy, becoming another cornerstone of its. Therefore, Pistorio coined the message—ecology is free—which was followed shortly after, precisely in 1995, by the publication of the first "Environmental Decalogue" a set of commitments taken by ST, towards several environmental fields such as: reduction of electricity, water, materials and chemicals consumption, reduction of air, water and noise pollution, reduction and recycling of waste, development of environmental management systems at all ST sites according to the dictates of the European Regulation EMAS recently published (1993); finally, compliance with the most stringent laws and regulations in the world and involvement of ST's stakeholders.

So began a long journey that brought all ST sites to structure and operate an environmental management system (EMS), being audited by an independent certification body. In the meantime (1996) was published the first version of the international standard ISO 14001. The Italian sites (Agrate Brianza, Cornaredo and Catania) arrived almost simultaneously at the finish. It was left to Catania, for a series of events that are worthless mentioning now, the pride to obtain the registration in December 1997 with the number IT—00001. The fifteenth anniversary of that important goal then befalls just in these days.

8.3 ST Microelectronics—Site of Catania. Historical Data and Environmental Performance

Catania manufacturing site of STMicroelectronics is located in the industrial area of Pantano d'Arci close to Catania. Coming from the south, on the National Road nr 114, you can see it just three kilometers far from the international airport "Fontanarossa".

The area was marshy until the fifties (hence derives its name), but was converted into agricultural land as a result of a massive reclamation carried out in those years. The first building was erected in 1961 by the USA company RCA, who sold it to ATES (Aziende Tecniche Elettroniche del Sud), which was in turn acquired by SGS Microelettronica.

The area is classified as industrial since the sixties; in the surroundings there are other industrial enterprises while there are not residential districts.

The factory occupies an area of about 18 ha of which about 25% covered by buildings dedicated to the production areas, the offices, the technical areas and the warehouses.

The site employs about 3,800 permanent units and includes two front-end production lines.

Since it has been just mentioned that the site implemented an environmental management system 15 years ago, what lessons can be learnt from such a long commitment to the environment?

First of all it is useful to take a look to the site's evolution, mainly to its production growth; in 1997 starts the new production line (8 inches wafers), which has been later on (2002) expanded, up to double its production capacity.

Such trend is clearly visible in the Fig. 8.1 that draws the production growth from 1997 to 2012. Can be noticed the light stop in 2001, and the significant drop in 2008–2009 in correspondence to the economic crisis that still continues, despite the good performance of 2010.

Let us now analyze the energy consumption in the same time-frame; looking at the absolute consumption (Fig. 8.2), we can note a gradual and almost continuous growth from 1997 to 2003, that is coincident with the production capacity increase. The consumption in the following years is almost flat and shows a fluctuation in a range not wider than 12% about.

Known these data, it's evident that the ratio between energy consumption and production volumes leads to obtain the specific consumption (Fig. 8.3) that appears continuously decreasing, particularly in the first years. The entire chart proves that the energy consumption per wafer produced, has been almost halved with only one exception; the financial crisis that began in the summer of 2007 had the worst peak at the end of 2008 and the countermeasures immediately deployed were not enough to mitigate the bad results in 2009.

The remarkable reduction in specific consumption of electricity was achieved through a series of improvements and systems modifications. The main are: the installation of VFD (Variable Frequency Drives commonly known also as Inverters)

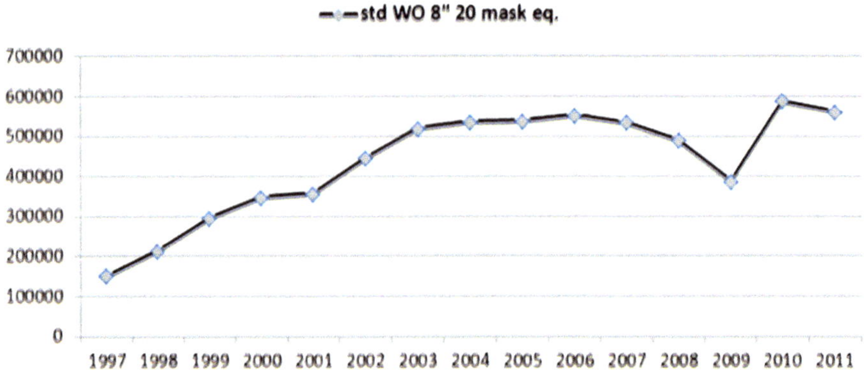

Fig. 8.1 Production trend in standard wafer (8 inches—20 masks equivalent)

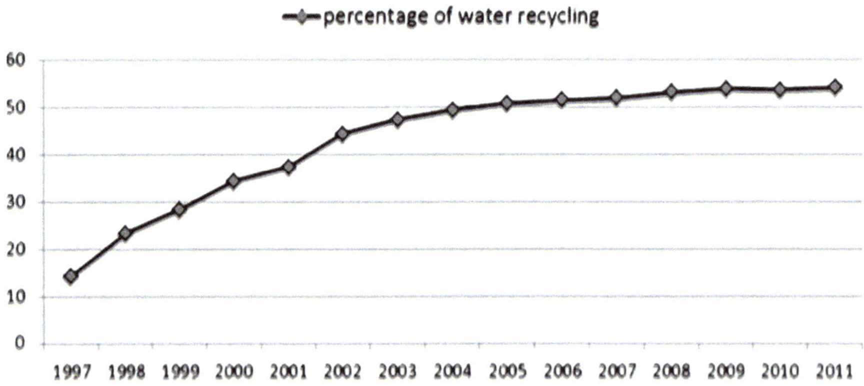

Fig. 8.2 Percentage of water recycling

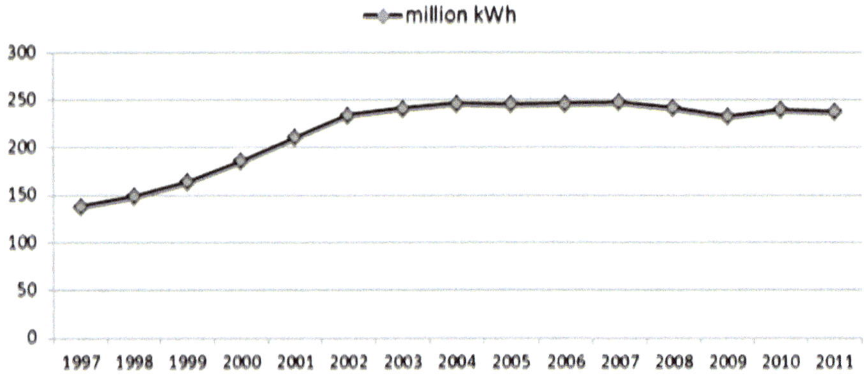

Fig. 8.3 Absolute consumption of electricity

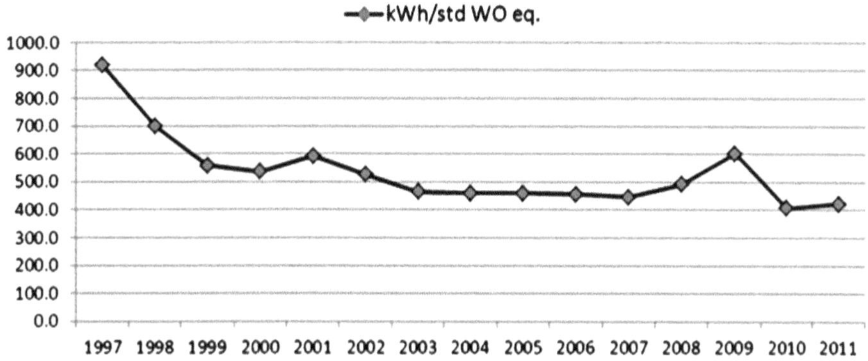

Fig. 8.4 Specific consumption of electricity (per wafer out equivalent)

in all those machines (pumps, fans, compressors, etc.) that need to adjust their motor's speed to closely match output requirements, the implementation of adiabatic cooling systems, the production of chilled water at different temperatures (5, 13 and 18 °C) to fulfill users' requirements, the replacement of old electric motors with new ones at high efficiency. Another crucial activity, the application of energy-saving measures e.g. setting accurately the lighting systems, sealing every leak in the compressed air pipework, shutting down totally some equipments/systems during the non-use period, etc. Last but not least, the erection of a photovoltaic system, with the support of an external partner, capable to generate approximately 1 % of the energy consumed by the site.

Another significant aspect to be considered is the water consumption trend (Fig. 8.4); increases in consumption precede the production growth because the startup of a new production line, and then of new machines that use ultrapure water, requires long testing and wasteful flushing cycles. In any case, in the period 1997–2003 we note a conspicuous progression, while in the following years, consumption becomes stable.

As per the electricity consumption, progress is evident when we look at the specific water consumption (Fig. 8.5); over the last 15 years they have been reduced to less than half.

Except for the year 2009, already mentioned for the low production volumes, the water consumption trend displays another singular point, in correspondence with the year 2001; tuning of processes in the new production area required more water without the relevant production of wafers.

The downward trend of specific consumption of water, besides some easy saving measures, is basically due to the recycling of the water (see Fig. 8.6); starting from a significant percentage of 14 % in 1997, it reached values higher than 50 %, from 2005 onwards.

This result has been achieved thanks to the erection of a system dedicated to the reclaim of water. It collects a mix of wastewaters with characteristics suitable for reuse. The appropriate blending of these waters combined with filtering and reverse

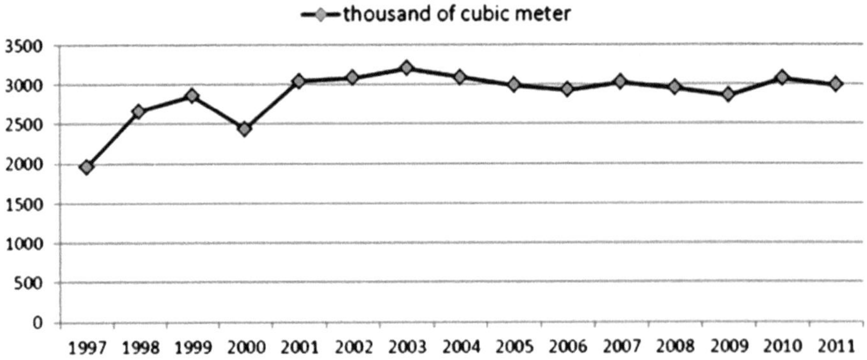

Fig. 8.5 Absolute consumption of water

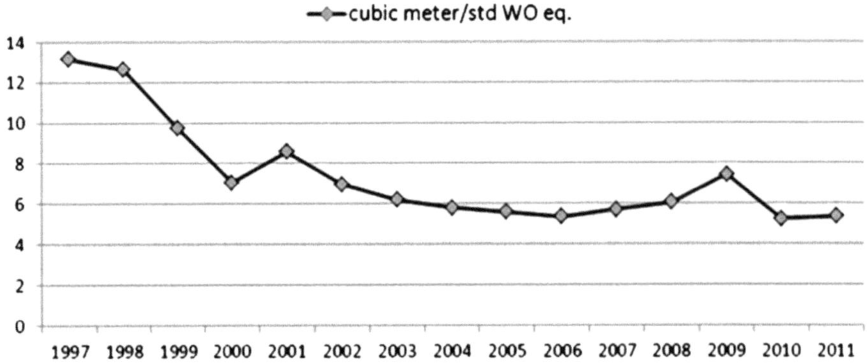

Fig. 8.6 Specific consumption of water (per wafer out equivalent)

osmosis treatment, make the final stream acceptable for industrial use (cooling towers, air emissions scrubbers, feeding of closed circuits, washings, etc.)

So far we have given only a brief mention of the results obtained in the environmental field by the ST site of Catania; following the strategy set by top management the factory sought other achievements in the fields of waste reduction and recycling, atmosphere and water emissions reduction, consumption of chemicals reduction, removal of hazardous materials such as asbestos, lead, etc.

The adoption of this management a system, gave the opportunity to deal with each problem in an orderly and methodical way, looking time by time the applicable regulatory constraints and the technological solutions most efficient and economically viable. Then have been defined clear objectives and specific action plans to achieve them. Also the environmental statement, prepared annually by the site, became a useful tool for collection, analysis and storage of data.

Despite a management system in an industrial enterprise operates efficiently, because it is well conceived and widely spread and used in all hierarchical levels of the organization, it is true that often it collides with an external reality not structured and organized as fine.

Therefore lack of adequate facilities such us incinerators or storage stations for special waste is common as well as lack of suitable wastewater treatment facilities or even the same public drainage network.

In other words, an industrial site in Italy has to solve, in a larger scale, the same problems that has every ordinary citizen, dealing with an "environmental" problem; in these cases, the response of the civil society must be systemic and organized to avoid any adverse effect on our territory.

8.4 Conclusions

At the end of this brief speech on environmental management systems, the most obvious conclusion is the expectation that this tool is adopted by the highest number of subjects, both private and public. A structured approach will in fact enhance the careful management of the territory, promote the search for processes and products with lower environmental impact, implement infrastructures and facilities technologically efficient and economically viable to manage environmental issues. Within the European Union, German companies have been the first that massively joined the European Regulation EMAS adopting its EMS. The current economic situation seems to validate the thought of Pasquale Pistorio and his words, "environmentally oriented industries are financially more competitive", after almost 20 years, sound as a prophecy.

References

Pieranunzi R (2001) TQM breeds success at STMicroelectronics. J Org Exc 20:39–43. doi:10.1002/npr.1105

Brown LR (2002) Eco economy. Editori Riuniti, Roma

Cianciullo A, Leonardi G (1996) Far soldi con l'ambiente. Sperling & Kupfer Editori, Milano

STMicroelectronics Environmental Statement of Catania Site (2000–2011). http://www.ctn.st.com/environment/index.htm. Accessed 5 Nov 2012

Chapter 9
Sustainability in the Food Oil Sector: The Experience of the *Oleifici Mataluni* in Eco-Innovation Projects

Elena De Marco, Maria Savarese, Antonella Ambrosone, Isabella D'Antuoni and Salvatore Falco

Abstract Environmental sustainability is becoming a factor for competitive advantage on the market, as the new-generation consumers are determined to take into consideration the environmental impact of their purchasing choices. Also food companies are giving attention to this market demand, introducing social and environmental responsibility into their policies. This chapter reports the experience of a company operating in the sector of food oils, as regards two projects aimed at introducing eco-innovation in the productive processes. The first innovation regards the introduction on the market of an eco-sustainable bottle for food oil, made up of 100% recycled plastic; the second innovation regards the application of a process for the valorization of an effluent resulting from the process of extraction of oil from olives, by recovering high-value bio-molecules and purified water, thus turning a potentially polluting waste into a resource. Both the project have received the support of financial instruments for the environment of the European Commission, aimed at promoting the diffusion of eco-innovation in Europe.

Keywords Industrial eco-innovations · CIP eco-innovation programme · Recycled PET bottle · LIFE + programme · Olive mill by-products valorization

9.1 Introduction

In our consumer-driven world, the potentially harmful impact of products and services on the environment has too often been ignored. But that is changing, thanks to a new generation of consumers determined to take into consideration the environmental impact of their choices.

E. De Marco (✉) · I. D'Antuoni · S. Falco
CRIOL, Research Centre for Olive Oil Industry–Industria Olearia Biagio Mataluni srl, 82016 Montesarchio, BN, Italy
e-mail: criol@mataluni.com

M. Savarese · A. Ambrosone
Fabio Mataluni & C. srl, 82016 Montesarchio, BN, Italy

R. Salomone, G. Saija (eds.), *Pathways to Environmental Sustainability*,
DOI 10.1007/978-3-319-03826-1_9,

According to a recent study by Nielsen, a leading global provider of information and insights into what consumers watch and buy, half of consumers under age 40 show willingness to pay extra for products and services from socially-responsible companies. Nielsen's Global Corporate Citizenship Survey of more than 28,000 Internet respondents in 56 countries shows that 46 % of global consumers are willing to pay extra for products and services from companies that have implemented programs to give back to society (Nielsen 2012).

This is why it is more and more important, for a company aiming at keeping its competitiveness on the market, to join, by means of industrial innovations, economic advantage for the enterprise and benefits for the environment.

The European Commission offers support to companies interested in developing and applying innovations with a positive impact on the environment, by means of some financial instruments.

The LIFE programme is the financial instrument for the environment of the European Commission. From its institution, in 1992, the LIFE programme has co-financed more than 3,100 projects sustaining environment across the EU, with an overall financial contribution of 2.2 billion euros. The LIFE programme has been articulated in five different stages. The current stage (LIFE+) started with EU Regulation 614/2007 and can manage a total budget of 2,143 billion euros for the period 2007–2013, during which annual calls for proposal are published by the European Commission. LIFE+provides financial support for projects that "contribute to the development, updating and implementation of Community environmental policy". The programme intends to fill the gap existing between the research and development activities and the application of their results on a real scale. The diffusion of innovations in environmental field is promoted by co-financing projects with a demonstration character, in which a key role is attributed to the diffusion of results and to divulgation activity.

Another EU financial instrument supporting the diffusion of products and services with positive impact on the environment is the programme CIP Eco-Innovation, which is specifically aimed at joining innovation, market and environmental sustainability. Supporting pilot projects, the initiative promotes the diffusion on the market of eco-innovative technologies, products and services, that could determine a better use of natural resources and reduce the European ecological footprint. Since the launch of the programme in 2008, almost 200 projects of first introduction on the market of eco-innovation products, involving more than 650 organisations in 5 sectors of activity have been selected.

This chapter describes the experience of an agro-food company making of eco-innovation a key factor for business and development.

Oleifici Mataluni are an industrial complex involved in the production and bottling of food oils and in the production of packaging for edible oils. Within their premises, the Oleifici Mataluni carry out the whole production process, from the olive oil milling and refining, to the oil bottling and packaging in glass, PET and can, to the production of PET bottles, caps, boxes and labels.

With the firm belief that the sustainability is a responsibility from which the companies cannot escape and that repays also in economic terms, Oleifici Mataluni have decided to invest in eco-innovation, in the sector of packaging, that plays a crucial role in determining the overall environmental impact of the food products, and in the sector of valorization of by-products of the production processes. In developing eco-innovations, Oleifici Mataluni have made recourse to the support of the European Union, exploiting the possibilities offered by the EU financial instruments for the environment.

9.2 Production of a Recycled Polyethylene terephthalate (R-PET) Bottle for Food Oil: The Project RE-PACK EDOILS

The improvement in the waste management is considered one of the greatest environmental challenges at international level. On a worldwide scale, the plan of implementation approved at the World Summit for Sustainable Development at Johannesburg in 2002, hopes for the implementation of actions aimed to "prevent and minimize waste and maximize reuse, recycling and use of environmentally friendly alternative materials".

Within EU, the Environmental Commission of the European Council highlights the necessity to break the link between economic growth, use of resources and waste production, underlining the need for economic growth to be harmonized with the rational use of natural resources and with a sustainable production of waste. To this aim, European environmental politics promote legislative actions aimed at a sustainable use of natural resources. Particular attention is given to the reduction of packaging waste, because of the high quantity of this material and of its short life cycle.

Among all packaging waste produced annually in Europe, those made of plastic materials amount to 10–12% by weight and 20–25% by volume and are thus responsible for a high environmental impact.

Thanks to the EU legislative actions, a significant increase of the quantity of recycled plastic material can be observed starting from 1990 up to now.

There is, thus, the possibility (and the necessity) to extend the commercial applications of recycled plastics to sectors that initially were precluded to them, such as food packaging. The strategy that seems to be the most rational and promising, in terms both of health security and of potential environmental and economic benefits, is the realisation of processes of "closed loop recycling", oriented to re-introduce re-manufactured products in the same sector from which the wastes are originated. During the last years, several schemes of closed loop recycling have been developed in order to recycle food bottles, as these products cover a wide segment in the packaging market, they can be easily identified and separated from other wastes and represent a high percentage by volume of plastic wastes.

The research work about methodologies of closed loop recycling and the resulting technological progresses have allowed to develop and certify modern recycling processes, able to produce R-PET with purity characteristics adequate to be

The project in summary

Title	Use of 100 per cent Post-Consumer Recycled Polyethylene Terephthalate to produce packaging for edible oils
Acronym	RE-PACK EDOILS
Agreement no.	ECO/10/277355/SI2.599775
Company	Fabio Mataluni & C. s.r.l., Montesarchio (BN), Italy
Starting date	09/11/2011
Duration	36 months
Investment	1,168,864 €
Allocated EU contribution	584,432 €

Fig. 9.1 The project RE-PACK EDOILS in summary

used in direct contact with food (super-clean recycling technologies), in accordance to the recent EU Regulation 282/2008.

This Regulation has introduced the possibility of using post-consumer recycled plastic materials (those deriving from the differentiated waste disposal) also for the packaging of food products, provided that the materials come from a certified recycling process and satisfy all the requirements foreseen for the food contact.

In line with this recent updating of the European legislation about the use of recycled plastics, the Oleifici Mataluni proposed, with the project RE-PACK EDOILS, the introduction of recycled plastics also in the packaging of edible oils.

The project received the financial support of the European Union within the CIP Eco-innovation programme (agreement no. ECO/10/277355), and was started up in November 2011 by the Fabio Mataluni & C., a company of the Oleifici Mataluni industrial complex involved in the production of packaging for edible oils. Some summary figures about the project are reported in Fig. 9.1.

Aim of the project is producing and introducing on the market packaging for edible oils made up of 100 % recycled plastics.

In the 3 years of activity, the project will lead to the production of bottles for food oils completely made up of post-consumer recycled polyethylene terephthalate (PCR-PET), and to their introduction on the market after a careful evaluation of their functional properties and their suitability for food contact (European Commission 2012).

Having already verified the technical feasibility of the idea, by means of preliminary experimental trials carried out by the corporate research centre, the

Fabio Mataluni & C. will start up a pilot line for the production of preforms, that will be converted in bottled for edible oils made up of 100 % recycled plastic. These bottles will be introduced on the market in the countries where the use of recycled plastic is admissible for food contact.

The use of recycled plastic proves to be highly advantageous from an environmental point of view, reducing the quantity of material to be sent to the rubbish dump and the dependency on non-renewable sources.

It was reported that the inclusion of recycled material in the PET bottle life cycle increases its global efficiency, in terms of emergy, in approximately 37 % sej/sej (Almeida et al. 2010). The concept of emergy can be used to assess the load imposed by a product to the environment. Emergy is defined as the quantity of solar energy necessary (directly or indirectly) to obtain a product (good or service) or energy flow in a given process (Odum 1996). Solar energy is the common basis of all energy flows circulating within the biosphere. The greater the emergy flow necessary to sustain a process, the greater the quantity of solar energy consumed or, in other words, the greater the environmental cost. Hence emergy represents the memory of all the solar energy consumed during the process. Emergy is measured in solar energy joules (sej). The implementation of the recycling stage reflects directly on the extraction and refining phases diminishing the quantities of resources and energy use. These data are referred to production of bottles with 40 % of recycled material in weight, while the project RE-PACK EDOILS aims to producing bottles made with 100 % of PCR-PET, so that the balance in this case should be much more profitable.

In a comparison of the overall environmental impact of three end-of-life management options for PET packaging (closed loop recycling, burning with heat and power recovery and burning with power generation), the recycling option resulted in an overall reduction in the emission of key pollutants and in the overall environmental impact (Chilton et al. 2010).

Considering the environmental advantages resulting from the use of recycled plastics in the production of food packaging, the company has decided, with the project RE-PACK EDOILS, to introduce on the market a new bottle for vegetable oils made up of post-consumer recycled PET, thus extending the use of recycled plastics to a new field of application.

9.3 Recovery of High-Value Natural Molecules from Olive Mill Effluents: The Project RE-WASTE

The agro-food sector, which plays an important role in the European economy, determines the production of great quantities of organic wastes, which management and disposal poses often serious problems.

Environmental legislation has significantly contributed to the introduction of sustainable waste management practices throughout the European Union, giving support to the development and diffusion of process for the valorisation and use of these wastes.

The olive mill wastewaters (OMWW) represent one of the most polluting and difficult to manage agro-industrial effluents in the Mediterranean basin, due to their high organic load and concentration of compounds recalcitrant to biological degradation, such as tannins, long chain fatty acids and phenolic compounds. For the treatment of OMWW several physical, chemical and biological processes have been proposed (Roig et al. 2006), such as evaporation in open-air lagoons, coagulation-flocculation, membrane separation (Paraskeva et al. 2007), anaerobic digestion (Hamdi 1996), composting (Tomati et al. 1995), aerobic treatment, adsorption (Soto et al. 2011), and other processes such as oxidation, thermal treatment and electrocoagulation (Paraskeva and Diamadopoulos 2006). Nevertheless, none of them proves to be a real solution for the OMWW disposal, feasible from a technical and economical point of view.

Currently the only solution that olive oil mill operators have at their disposal for the managing of olive mill wastewaters is the agronomic use, i.e. spreading on agricultural soil, according to the Italian law 574/1966. Nevertheless, the prescriptions imposed (choice of the place suitable as regards the morphology and the geology, maximum limit of effluent per hectare, prohibition of spreading on a soil saturated in water, …) make it difficult for the olive oil operators to spread the effluent respecting the law.

As a consequence, there is an urgent need of alternative solutions for the managing of olive mill effluents, by means of technologies that reduce their environmental impact and lead to a sustainable use of the natural resources.

The OMWW is characterized by a high concentration in phenolic compounds, which on one hand have a negative effect on the biological treatment as they inhibit the microbiological growth, on the other hand possess important biological properties, such as antioxidant, anti-inflammatory, anti-viral and anti-carcinogenic activities (Obied et al. 2005).

To recover, from the olive mill effluents, natural antioxidants to be used in the cosmetic, food or animal feed industry will allow to convert a problematic effluent into an industrial resource to valorize.

Industria Olearia Biagio Mataluni srl, a company of the Oleifici Mataluni industrial complex involved in the production and processing of olive oil and other vegetable oils, has been carrying out research activities aimed and finding a sustainable solution for the management and valorization of olive mill effluents.

With the support of the European Commission within the LIFE+ programme, Industria Olearia Biagio Mataluni has carried out a demonstrative project aimed at showing to the olive oil sector an innovative process for the valorization of OMWW.

The project RE-WASTE (Recovery, recycling, resource. Valorization of olive mill effluents by recovering high added value bio-products), co-funded by the European Commission within the LIFE+programme (project no. LIFE07 ENV/IT/421), has allowed Industria Olearia Biagio Mataluni, in collaboration with other three project partners (Euroimpresa SpA, Parco Scientifico e Tecnologico di Salerno e delle Aree Interne della Campania, and the Spanish Centro Tecnológico Nacional de la Conserva y Alimentación) to test the process at demonstrative scale, showing the result to the European olive oil sector.

A semi-industrial automated plant for the treatment of OMWW was installed next to the olive oil mill of Industria Olearia Biagio Mataluni srl. The system is based on membrane filtration technologies (ultrafiltration, nanofiltration, reverse osmosis), preceded by a pre-treatment with chemical-physical processes (flocculation, centrifugation, filtration) and followed by a purification of the phenolic extracts on adsorbent resins. The scheme of the process is shown in Fig. 9.2.

Several works have used membrane filtration to separate from OMWW the phenolic fraction to be recovered (Russo 2007; Garcia-Castello et al. 2010; Paraskeva et al. 2007). In these studies the membrane filtration process was applied using only laboratory units or little pilot units. None of them applied the process at a semi-industrial scale. Also adsorption has been proposed as a valuable technique to recover, concentrate and purify phenolic compounds from industrial wastewaters (Soto et al. 2011), due to its relative simplicity of design, operation and scale up, high capacity, favorable rate and ease or regeneration. Additionally, it avoids using toxic solvents and minimizes degradation. Literature is available on the adsorption of phenolics on synthetic polymeric adsorbents and on the application of adsorption for the removal of polyphenols from OMWW (Agalias et al. 2007).

In the RE-WASTE project, a combination of membrane filtration and of adsorption techniques is used for the treatment of OMWW, at a significant demonstrative scale.

The proposed technology proved able to turn a polluting residue into a valuable source of molecules with biological activities and purified water by means of an environmentally sustainable process.

The treatment tested within the project combines different technologies and proved effective in recovering from OMWW a high quantity (65%) of purified water which could be re-used in industrial processes (such as washing water, for example) and a concentrate with antioxidant properties rich in hydroxytyrosol and other phenols. This extract could be used for producing functional foods, cosmetics or nutraceuticals.

9.4 Conclusions

Environmental sustainability is becoming a factor for competitive advantage in the food industry supply chain, as consumer preferences are evolving towards the demand for eco-products. Food industries have started adopting sustainability as a competitive advantage. Key stakeholders are significantly increasing pressure within the marketplace to encourage companies to adopt emerging environmental and recycling-oriented values into their commercial, consumer and policy choices.

Considering this evolution of the market, Oleifici Mataluni has decided to choose environmental sustainability as one of the key values of its policy. To this aim, it has oriented its innovation programmes towards the improvement of the environmental profile of its products, by reducing the environmental impact of the packaging, on one hand, and by finding solutions for the valorization of by-products and wastes of the industrial processes, on the other hand.

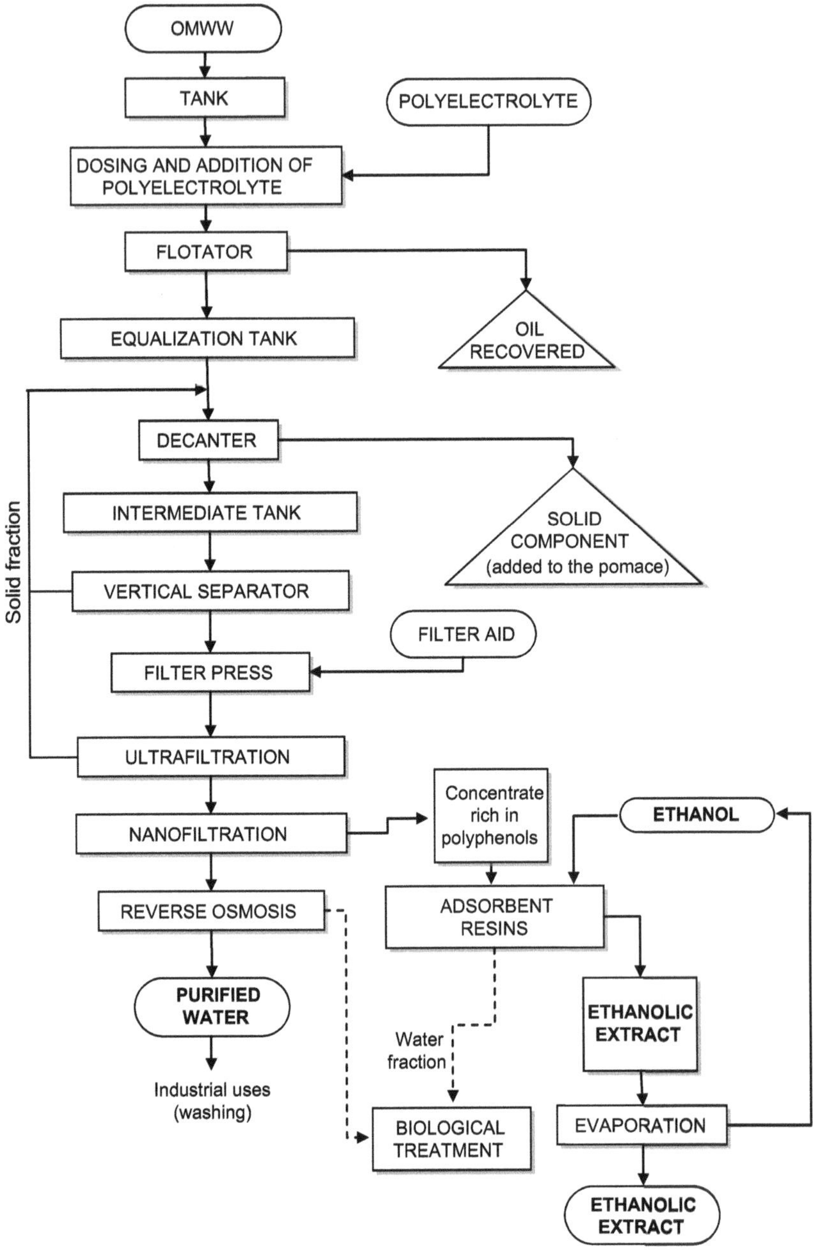

Fig. 9.2 The scheme of the process applied within the RE-WASTE project

The financial instruments of the European Commission, promoting the introduction on the market of eco-innovations and the diffusion of environmental friendly products and processes, proved very useful in supporting the efforts of the company in improving its environmental profile.

References

Agalias A, Magiatis P, Skaltsounis AL, Mikros E, Tsarbopoulos A, Gikas E, Spanos I, Manios T (2007) A new process for the management of olive oil mill waste water and recovery of natural antioxidants. J Agric Food Chem 55:2671–2676

Almeida CMVB, Rodrigues AJM, Bonilla SH, Giannetti BF (2010) Emergy as a tool for Ecodesign: evaluating materials selection for beverage packages in Brazil. J Clean Prod 18:32–43

Chilton T, Burnley S, Nesaratnam S (2010) A life cycle assessment of the closed-loop recycling and thermal recovery of post-consumer PET. Res Cons Recycl 54:1241–1249

European Commission (2012) Database of Eco-innovation projects. RE-PACK EDOILS project information sheet. http://eaci-projects.eu/eco/page/Page.jsp?op = project_detail & prid = 2101

Garcia-Castello E, Cassano A, Criscuoli A, Conidi C, Drioli E (2010) Recovery and concentration of polyphenols from olive mill wastewaters by integrated membrane system. Water Res 44:3883–3892

Hamdi M (1996) Anaerobic digestion of olive mill wastewaters. Process Biochem 31:105–110.

Nielsen (2012) The Global, Socially Conscious Consumer. http://www.nielsen.com/us/en/insights/reports-downloads/2012/the-global–socially-conscious-consumer.html. March 2012

Obied HK, Allen MS, Bedgood DR, Prenzler PD, Robards K, Stockmann R (2005) Bioactivity and analysis of biophenols recovered from olive mill waste. J Agric Food Chem 53:823–837

Odum HT (1996) Environmental accounting, emergy and decision Making. Wiley, NY

Paraskeva P, Diamadopoulos E (2006) Technologies for olive mill wastewater (OMW) treatment: a review. J Chem Technol Biotechnol 81:1475–1485

Paraskeva CA, Papadakis VG, Kanellopoulou DG, Koutsoukos PG, Angelopoulos KC (2007) Membrane filtration of olive mill wastewater (OMW) and OMW fractions exploitation. Water Environ Res 79:421–429

Roig A, Cayuela ML, Sánchez-Monedero MA (2006) An overview on olive mill wastes and their valorisation methods. Waste Manage 26:960–969

Russo C (2007) A new membrane process for the selective fractionation and total recovery of polyphenols, water and organic substances from vegetation waters (VW). J Membr Sci 288:239–246

Soto ML, Moure A, Domínguez H, Parajó JC (2011) Recovery, concentration and purification of phenolic compounds by adsorption: a review. J Food Eng 105:1–27

Tomati U, Galli E, Pasetti L, Volterra E (1995) Bioremediation of olive mill wastewaters by composting. Waste Manage Res 13:509–518

Chapter 10
Waste Management and Environmental Impact: A Case Study of Pharmaceutical Industry

Ayse Guney Agar, Gabriella Arcese and Maria Claudia Lucchetti

Abstract The European Initiatives of environmental policy are moving increasingly towards the continuous improvement of performance in a sustainable model, leading to stipulate that one of the components that quantifies the quality of the product on key variables of sustainability influences. The greater environmental impact in many manufacturing stages is the waste production and management. In this context, the life cycle approach better identifies and quantifying material and energy consumption and environmental emissions. The aim of this chapter is to describe, in detail, production and waste management in the pharmaceutical context, analyzing the environmental impact through life cycle approach.

Keywords Waste management · Life cycle approach · Environmental impacts · Pharmaceutical industry · Waste cycle · Classification criteria for substances

10.1 Introduction

In all societies, waste management and in particular industrial waste, creates a problem not only in terms of quantity of waste produced, but also problems faced in terms of various production cycles. Modern waste management practices are inspired by the principles of sustainable development. Thanks to environmental management, it is economically proven that implementing waste management and recovering the most of the raw materials where they were seen lost before, during the production stage, have now become a common sense that they have a great impact on revenue. Nowadays, we can allocate the least possible amount of landfill waste thanks to more modern systems environmental management.

Although pharmaceutical companies recognize the importance of including the environmental aspects from the beginning of product design process, and some of them have looked at concepts such as green chemistry and atom economy. In the

M. C. Lucchetti (✉) · A. G. Agar · G. Arcese
Department of Business Studies, Università di Roma Tre, 00154 Rome, Italy
e-mail: mariaclaudia.lucchetti@uniroma3.it

R. Salomone, G. Saija (eds.), *Pathways to Environmental Sustainability*,
DOI 10.1007/978-3-319-03826-1_10,

past, the application of this model has been limited to processing in chemical or fermentation. Source reduction of wastes can be achieved in industry through changing in products, raw materials, process technologies or procedural practices and through the reengineering of the life cycle in a sustainable pattern.

There is a need to introduce environmental factors as an integral part of the decision-making process at the Research and Development (R&D) stages of design for drug manufacturing. It is necessary to improve general efficiency of the company to lower the costs, and avoid the environmental consequences in the production stage (Arcese et al. 2012).

Recent international organization studies on environment show that pollution from the pharmaceutical industry is an element of concern on the potential environmental consequences that may result. The EMEA (The European Medicine Agency Science Medicines Health) has recently proposed some guidelines for assessing environmental risks of new drugs prior to registration, while other proposals coming from the so-called independent "green pharmacy". The international community through organizations such as the *United Nations Environment Programme* (UNEP)/Society of Environmental Toxicology and Chemistry SETAC (2010) is pursuing initiatives (Life Cycle Initiative), whose mission is to spread the life-cycle approach in the world (Komninos 2002).

10.2 Materials and Methods

For this study, we have utilized national and International publications, material online and material distributed on various conferences on environmental issues. Specifically on waste from pharmaceuticals industry, we have used personal conversations, interviews and published material from web, and internal technical documentation for the data analysis (Arcese et al. 2009).

Finally, to have a broader and more comprehensive overview, we have confronted this study with the guidelines which is provided by Confindustria and Federchimica, AFI (Associazione Farmaceutici Industria) and Certiquality (Certiquality et al. 2008). As methodology, in this chapter, Life Cycle Approach is used.

10.2.1 Pharmaceutical Industry and Environmental Impacts

Pharmaceuticals and Personal Care Products as Pollutants (PPCPs) refers, in general, to any product used by individuals for personal health or cosmetic reasons or used by agribusiness to enhance growth or health of livestock (Daughton and Ternes 1999).

PPCPs comprise a diverse collection of thousands of chemical substances, including prescription and over-the-counter therapeutic drugs, veterinary drugs, fragrances, and cosmetics (Daughton 2009; Daughton and Ternes 1999).

Their effect on the environment is now recognized as an important area of research. Research has shown that PPCPs are present in water bodies throughout the world. While some studies have suggested that these substances cause ecological harm, no studies have shown a direct impact on human health. More research is needed to determine the effects on humans of long-term exposure to low levels of PPCPs (Daughton 2009).

Moreover, all effects of low concentrations mixtures effects in different PPCPs are also unknown. While all effects of most PPCPs on the environment are not well understood, there is a concern about the potential harm as they may act unpredictably when mixed with other chemicals and might leak to the environment or into the food chain. In addition, some PPCPS are active at very low concentrations, and they are often released continuously in large or widespread quantities (den Boer et al. 2007).

Individuals release PPCPs to the environment through excretion (the elimination of waste material from the body) and disposal of unwanted medications to sewers and trash. In February 2007, the White House Office of National Drug Control Policy issued the first consumer guidance for the Proper Disposal of Prescription Drugs. Proper disposal of drugs is a straightforward way for individuals to prevent pollution (Daughton 2009).

Some PPCPs are easily broken down and processed by the human body or degrade quickly in the environment, but others are not easily broken down and processed, so they enter into the domestic sewers. Excretion of biologically unused and unprocessed drugs depends on individual drug composition or ability of individual bodies to break down drugs. (den Boer et al. 2007).

10.2.1.1 Classification Criteria for Substances in Pharmaceutical Industries

The waste management in pharmaceutical sector, in particular in the disposal phase, requires even more advanced technological solutions. For a substantial part of waste, the thermo-destruction is a necessary step for legal requirements because it is the best technology capable of reducing the risks of harming the environment (Caravita 2001).

The remaining part of the waste is derived from the processes of production or from residues of working. For this reason, it can be classified as special and similar waste according to the characteristics of the waste generator.

For the particularity of the substances composition in question, during storage, hazardous waste removal it is absolutely prohibited through sewers, municipal solid waste or emissions.

Each chemical laboratory or a company that carries out such activities should be equipped with special containers for the collection of exclusive special waste and hazardous waste. Since the entry into force of the Annex to Decision codes, CER derived EEC/EAEC/CZECH 118 of 16.01.2001 have been listed as hazardous waste and assigned a special code marking them with an asterisk and the same code will be reported in the discharge of the waste container.

Table 10.1 European Classification on waste categories

Similar municipal wastes	Special wastes	Hazardous waste	Packaging
Paper CER 150101	Expired medication CER 070514	Health risk of infection CER 180103*	Aluminium CER 150104
Glass CER 200102	Bottom ash CER 100101	Residues filtration and absorbent materials, halogenated CER 070509*	Paper CER 150101
Wood CER 150103		Coal fly ash CER 190402*	Plastic CER 150102
Printer toner CER 150106		Used oil CER 130203*	Iron/Steel CER 170405
Plastic cans wet fraction CER150106		Led CER 160601*	

Note: Asterisk * is the code for the Hazardous Waste CER (Waste European Code)

The company needs to be very careful not to mix substances in the same container which might react with one another and it might turn into harmful vapor or cause explosions. Mixing is not prohibited in an absolute, so before performing mixing of any waste products, the "MSDS" of the product should be verified. Containers must be made of: polyethylene and polypropylene, with wide mouth, provided with sealing plugs, and have a volume 5 l or less. Containers must be kept at the place of production, in safe places or in a ventilated deposit which exclusively should be used for storage purposes, and away from heat sources. As shown in Table 10.1, from European Classification on waste categories is possible to catalog, through CER code, wastes that are generated from the production process (Arcese et al. 2011).

During disposal, it is very important that waste hazardous containers labeled with the title of the department of origin, the name of the manager of the laboratory, the contents with the respective volume of each component, the EWC, the hazard class, the date of the transfer to temporary storage and the signature of the Head of the Laboratory (Bootman and Townsend 1993; Florence and Attwood 2002).

Cataloging is made based on the rejection predominant in case of mixing different chemical compounds. For safety reasons, the card which describes the characteristics must be delivered to the person responsible in laboratories who is also in charge of waste disposal and hazardous chemical containers that are placed in a temporary storage. In addition, the person in charge is the same person who will arrange an authorized firm for disposal in this specific activity. In Italy, similar to medical waste (ex ROT, Italian acronym of processed hospital waste), the Legislative Decree 219/2000 regulates the hospital waste requires the same solutions as above. We can define as hospital waste: disposable material usually contaminated by organic liquids or materials, cutting tools, experimental animals and livestock waste, pharmaceutical waste and containers. Biological agents such as genetically modified organism, cell cultures, human end parasites which may cause infection, allergy, poisoning, are classified according to criteria of dangerousness into four groups:

- Group 1: agents with no or low individual and community risk;
- Group 2: moderate individual risk, limited community risk;

- Group 3: high individual and low community risk;
- Group 4: high individual and community risk.

10.3 Empirical Analysis: A Case Study of an Italian Pharmaceutical Company

Management through the "integrated waste cycle" requires, in addition to the collection, sorting, recovery or disposal of waste, in various possible ways, a series of operations to exert, and coordinate with each other for environmental protection and in accordance with local regulations. At present, we can count on the following treatment systems: collection, recycling and recovery of materials and energy, incineration and land filling.

The quantity of waste sent for recovery is increasing over the years, in particular with regard to the activities of recycling/recovery of other inorganic substances, largely arising from the activities of construction and demolition, subjected to treatment in most cases in plants crushing or used in environmental restoration and processes related to the construction or road works. For disposal operations, in absolute terms there was an increase relates mainly to chemical-physical and biological treatment. It is a form of management in this process of expansion. The use of landfill remains the most used with a total number of plants in operation throughout the national territory amounting to 557 plants, which is, nevertheless a decrease over the previous year. The waste incinerated is in most cases made of special hazardous waste biomass. In quantitative terms, in Italy, approximately 520,000 t of hazardous waste, or about 46 % of the total are incinerated. It should be borne in mind that the spread of this practice for this category is due to the fact that, for certain specific categories, such as medical waste, the legislation specifies this treatment as a priority.

Proper waste management requires the rational use of these systems, with a trend towards overcoming the landfill. In the company of the case study, faced with a total production of 107.5 million t of annual waste products, the total that is managed approximately 101.6 million t, broken down as a percentage in 93 % of non-hazardous waste and 7 % of a dangerous nature.

The most prevalent form of special waste management is represented by the recovery of raw materials (49.4 million t) while the rest is destined for disposal (35.6 million t). A quantity of 13.9 million t remaining from the total of 101.6 million t of waste is opened prior to the filing or placed in reserve.

For all the waste managed which can be decomposed into: waste disposed of in landfills (19.4 %), waste sent to biological treatment (14.6 %), and finally the amount of waste incinerated (1.1 %).

The pharmaceutical industry waste has different nature. Residues of production cycles, production waste, liquid and gaseous waste from industrial machines in the drug companies are treated similar to medical waste and expired drugs and they are withdrawn from the market. The operational phase of waste management within a company can be summarized in Fig. 10.1.

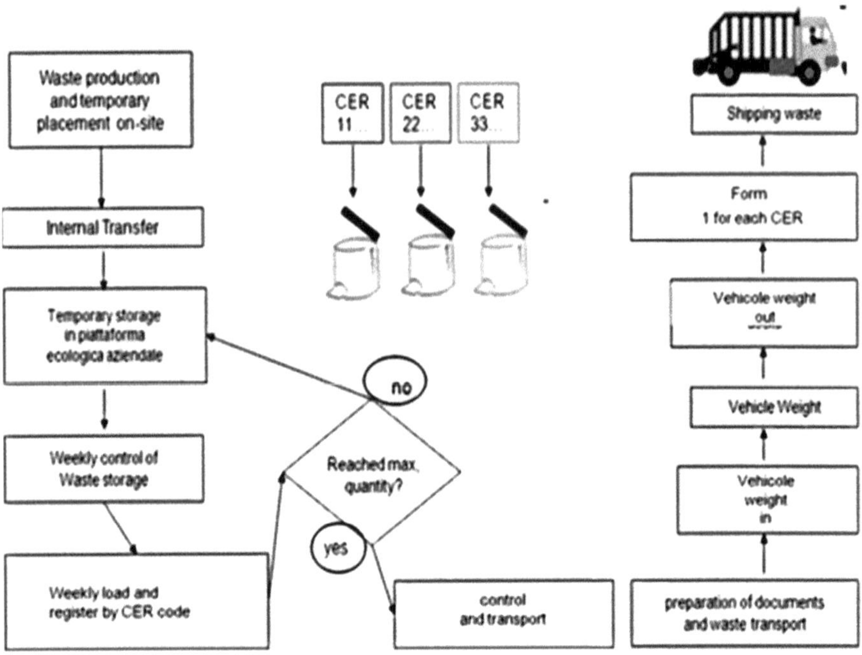

Fig. 10.1 Operations Management for waste management in pharmaceutical company. Based on guideline of Federchimica

10.3.1 The Life Cycle Approach in Pharmaceutical Production

From years, Life cycle approach is recognized in industry as a tool of particular importance for the evaluation of environmental impacts.

In 1990s, LCA (life cycle assessment) in pharmaceutical production process, selection and development were a novel area of research for improving process development in pharmaceuticals, increasing use of techniques such as combinatorial chemistry and discovery chemist to develop several chemical routes for the synthesis of a particular drug (EPA et al. 1987; EPA 2004).

The green chemistry and the necessity of measuring the sustainability objectives have led to increasing application of LCA in the chemistry and drug industry. A possible advantage is the possibility of explaining the production through a holistic, flexible and multidisciplinary tool (ISO 2006).

It enables to identify the critical hot spots and hierarchically classify processes with successful integration between technical assessment and economic evaluation. However, the problems of application are numerous: allocation problem (in particular because the productions are often multiple), difficulties in defining the boundaries of the system, lack of data.

Following the logic of life cycle thinking, in a cradle to-gate approach, the boundaries of the life cycle assessment are set to analyze the activity from the ex-

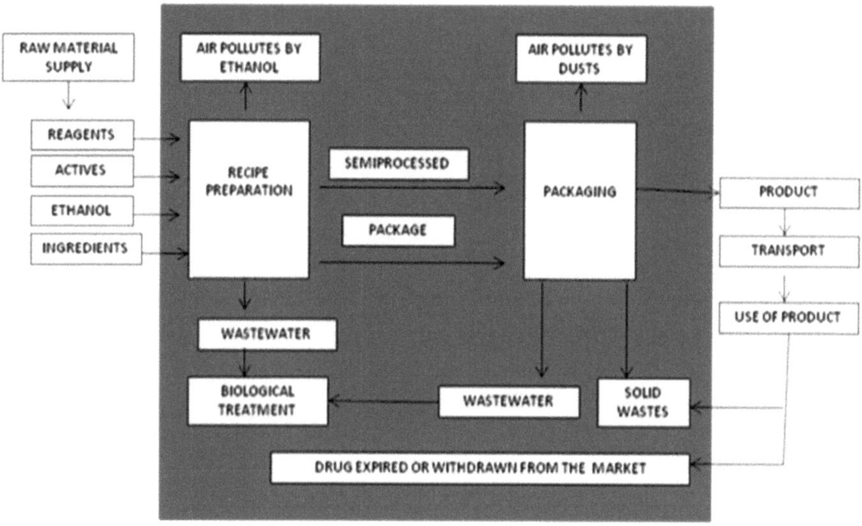

Fig. 10.2 Flow chart of pharmaceutical production system, our elaboration. (Arcese et al. 2011)

traction of raw materials until the end of the manufacturing process. The flow chart represents the system object of the analysis.

In general, in a pharmaceutical production process of a drug is substantiated in two main phases: the preparation of the recipes in the laboratory and packaging of the drug. Being simple compounds, mainly directed to treat minor elements and short-term, affecting the health transiently and are easily recognizable by the common experience of the patient (Fig. 10.2).

10.3.1.1 The Production—Mixing Stage

The location of chemicals, substances and preparations in a pharmaceutical laboratory are provided by the guidelines which is still a rare application. The waste in question, prior to exit from a single laboratory, must be harmless under the responsibility of the producer who only has the right mix hazardous materials (Notarnicola 2008; Official Gazette of Italian Republic 1997).

10.3.1.2 The Packaging Stage

During the sterilization, the packaging material is made of plastic and the covering film prevents contamination from the environment. To ensure aseptic conditions for dependability, the dispenser shall be equipped with special devices and follow a series of pre-sterilized materials. Sterile implants are placed outside of the machine, but an integral part of the packaging stage which is composed of valves, filters,

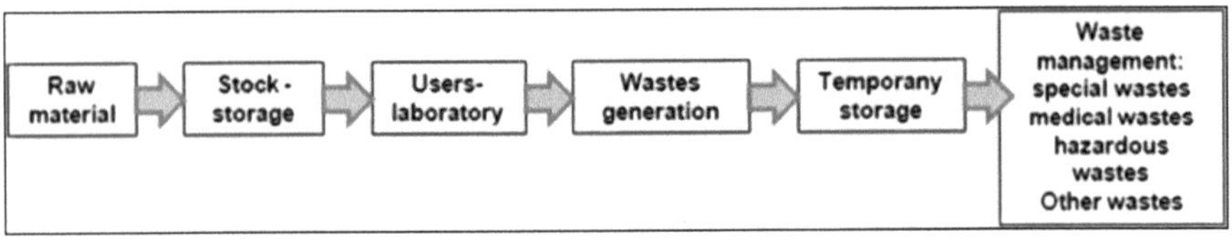

Fig. 10.3 Flow chart of waste production

devices and circuitry, and directs the sequences and cycles of the machine is pre-sterilized during the production phase.

10.3.2 Waste Cycle in the Pharmaceutical Industry: Results

The life cycle of a pharmaceutical product begins with the discovery of an active substance characterized by therapeutic potential. Before the product is launched on the market, testing phases, efficacy and safety stages of the product are necessary and they are very long. These stages include periods of varying duration, on average 12 years.

The company owner has an interest in applying innovation, patent protection since the discovery of the active ingredient, though, so it was, the actual time duration of the patent is inevitably limited. In fact, the phases of research and development after the initial discovery erode the actual time in which the product has marketing exclusivity, and this situation has encouraged the industry to minimize the time of placing the product on the market (time to market) (European Union Commission 2008).

The life cycle of waste in the pharmaceutical industry follows the flow chart shown in Fig. 10.3.

The specific amount of solid waste production is mainly the input to the packaging; in fact, the largest quantity is composed by paper and glass.

Furthermore, with the greatest impact are the waste for which the Italian Law requires a specific final treatment (special medical waste incinerators and landfill for hazardous waste, used oil and led are run by consortia required). In Fig. 10.4, different waste listed is shown in tons. The current average percentage of waste destined for recycling in this company is 60 %, which is similar to urban waste and could totally be recycled. The specific amount of solid waste production is mainly input to the packaging. The largest quantities of materials is paper. Around 40 % of glass and paper used for packaging is recycled while the remaining part is burned and the ash is placed in waste disposals.

The greatest impact is the waste for which the Italian law requires a specific final treatment; for example, medical waste incinerators and landfill for hazardous waste. Used oil and led are run by consortia required.

Around 40 % of glass and the paper used for packaging are recycled while the remaining part is burned and the ash is placed in waste disposals. The tradition-

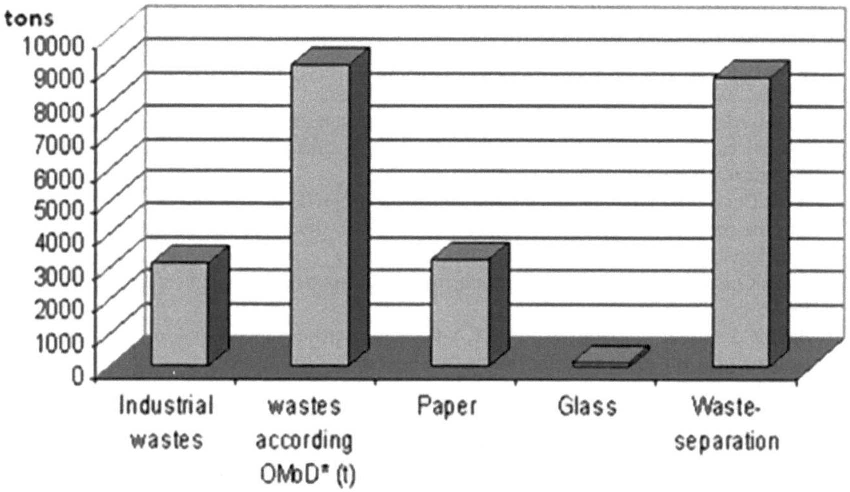

Fig. 10.4 Quantitative allocation of waste production. *Ordinance on the movement of Waste Code. The OMoD defines waste categories

ally more impactful phases are packaging and transport (which in this study were excluded from the boundaries of the system) and the analysis reflects these results.

There is no doubt of the importance of the evaluation of a production system and in particular environmental sustainability of pharmaceutical products. In general, the environmental analysis in this sector shows that the main effluents of the sector, beyond the use of fossil fuel, are wastewater and organic solid waste. Problems associated with waste generation in the wine industry are of special relevance during packaging phase. However, in this sector the use of large amounts of chemical substances and organic matter generates the possibility to recover organic waste. The LCA method is a valid tool to study environmental systems behavior. The use of renewable resources, sunlight and wind, could be a solution to decreasing fossil fuel consumption. All this would result in a significant decrease of economic cost and rationalization of resources.

References

Arcese G, Lucchetti MC, Martucci O (2009) Gestione dei rifiuti nella industria farmaceutica: una ricerca empirica. In: Percivale F, Varese E, Arese T (eds) Ambiente, Internazionalizzazione, Sistemi, Merci, Energia—Proceeding of 24th National Congress of Commodity Science, vol 1, Celid, Turin

Arcese G, Lucchetti MC, Martucci O (2011) The importance of Life Cycle Assessment for environmental sustainability: an empirical study of a pharmaceutical product. In: Koroneos C, Rovas DC, Dompro AT (eds) Proceeding of Elcas 2011–2nd International Exergy, Life Cycle Assessment, and Sustainability Workshop & Symposium. Ziti Publications, Thessaloniki, Greece

Arcese G, Lucchetti MC, Martucci O (2012) Analysis of sustainability based on Life Cycle Assessment: an empirical study of wine production. J Environ Sci Eng 5B(1):682–689

Bateman I, Kerry Turner R, Pearce DW (1994) Economia ambientale. Il Mulino, Bologna
Bootman JL, Townsend RJ (1993) Introduzione alla farmaco economia. Agenzia Italiana del farmaco, Milan
Caravita B (2001) Diritto dell'ambiente. Il Mulino, Bologna
Certiquality, AFI, Federchimica (2008) Linee guida per la gestione delle attività operative relative ai rifiuti industriali (Settori chimico-farmaceutico, cosmetico, biotecnologico, omeopatico e integratori alimentari). Milan
Daughton CG (2009) Pharmaceuticals and personal care products in the environment, chemicals from the practice of healthcare: Chalenfes and unknowns posed by residues in the environment. Environ Toxicol Chem 28(12):2490–2494
Daughton CG, Ternes TA (1999) Environmental health perspectives, annual report, 107. EPA, USA
den Boer J, den Boer E, Jager J (2007) LCA-IWM: a decision support tool for sustainability assessment of waste management systems. Waste Manag 27:1032–1045
EPA (2004) Guidelines for water reuse. Washington DC
EPA, FEMA, US Department of Transportation (1987) Technical guidance for hazard analysis—emergency planning for extremely hazardous substances. USA
European Union Commission (2008) Comunicazione della Commissione al Parlamento Europeo, al Consiglio, al Comitato Economico e Sociale Europeo e la Comitato delle Regioni. Due volte 20 per il 2020. L'opportunità del cambiamento climatico in Europa. http://europa.eu/legislation_summaries/energy/european_energy_policy/l28203_it.htm. Accessed 23 Jan 2008
Florence AT, Attwood D (2002) Tecnologia farmaceutica. Edises, Naples
ISO (2006) ISO 14040 'Environmental management—Life cycle assessment—Principles and framework'. International Organisation for Standardisation, Geneva
Komninos I (2002) Product life cycle assessment management. URENIO—Urban and Regional Innovation Research Unit, Thessaloniki. http://www.urenio.org. Accessed 23 Jan 2008
Notarnicola B (2008) Strumenti tecnici a supporto delle certificazioni ambientali: l'Analisi del Ciclo di Vita (LCA). In: Buonfrate A (ed) Codice dell'Ambiente. UTET, Turin
Official Gazette of Italian Republic (1997) Legislative Decree 22

Chapter 11
Subcategory Assessment Method for Social LCA: A First Application on the Wine Sector

Paola Karina Sanchez Ramirez, Luigia Petti and Cassia Maria Lie Ugaya

Abstract This chapter presents a Subcategory Assessment Method (SAM) which aims to support and normalize the Social Life Cycle Assessment (S-LCA) case studies regarding subcategory assessment. The method enables the social profile of the organizations involved in the processes along the product life cycle, using a four-level scale for each subcategory. SAM was applied in a small winery; it showed that for most of the subcategories the company did not reach the basic requirement. However, in some cases, simple changes could improve their performance. The results of the case study showed that it was possible to collect data and to evaluate the company using SAM. S-LCA is as time and work demanding as (environmental) LCA. SAM could be also implemented for the entire product life cycle. Future development of SAM could be the method expansion to other stakeholders and subcategories listed in the UNEP and SETAC (2009) guidelines, adapting the basic requirement for each of them.

Keywords Social life cycle assessment · Subcategory assessment method · Wine · Stakeholders · Product social profile

11.1 Introduction

Social Life Cycle Assessment (S-LCA) is a technique used to evaluate potential positive and negative impacts along the life cycle of a product (UNEP and SETAC 2009). S-LCA is quite new, assesses the behavior of organizations and evaluates their capacity for creating and developing a relationship with the various stakeholders.

P. K. S. Ramirez (✉) · L. Petti
DEC—Department of Economic Studies, G. D'Annunzio University, Viale della Pineta, 4, Pescara, Italy
e-mail: p.sanchez@unich.it

C. M. L. Ugaya
PPGEM—Post Graduation Program of Materials and Mechanical Engineering, Federal Technological University of Parana, Avenida Sete de Setembro, 3165, 80230-90, Curitiba, Brazil

R. Salomone, G. Saija (eds.), *Pathways to Environmental Sustainability*,
DOI 10.1007/978-3-319-03826-1_11,

S-LCA follows four phases, similarly to an environmental Life Cycle Assessment (LCA): goal and scope definition, inventory analysis, impact assessment and interpretation; nevertheless it requires adaptations (Weidema 2005).

The transition from raw data to impact assessment in S-LCA is still a challenge and can be performed in two ways (Type 1 and Type 2; UNEP and SETAC 2009). Type 1 assigns the Inventory Indicators (II) to 31 subcategories related to five stakeholders: employees, local community, society, consumers and other actors in the value chain (UNEP and SETAC 2009). In this case, the II provide information about a specific subcategory, for example, the contractual agreements related to overtime is a II related to the subcategory working hours of the worker stakeholder. Type 2 assigns the II to social impacts through social impact pathways, similar to environmental LCA.

Over the years, some methods have been presented in the literature for the characterization step. The first proposal for Type 2 was developed by Weidema (2006) who created a new indicator at endpoint level, the Quality Adjusted Life Years (QALY). This indicator is composed by six damage subcategories under the general category "human well-being". QALY aggregates the results by a combination of different statistical data. QALY should be understood in a similar way as DALY of World Health Organization (WHO). Therefore, QALY represents the reduced quality of living by shortening the life expectancy. Norris (2006) related national gross product with life expectancy of several countries and developed a characterization factor that could be used to wages. In the same period, Hunkeler (2006) suggested the use of labor hours as an intermediate variable in the calculation for societal life cycle assessment. He assumes that processes can be dismembered into labor statistics, considering only a single impact category.

Life Cycle Impact Assessment methods related to Type 1 have already been proposed (Dreyer et al. 2006; Ciroth and Franze 2009). The first one was the leading work in consider subcategories, but limited only to worker stakeholder. The second one (Ciroth and Franze 2009) proposed a method in 2009 and improved it in 2011 (Ciroth and Franze 2011), which includes all stakeholders and subcategories, without, however, establishing an objective way to evaluate data to the subcategory and how to aggregate subcategories into impact categories.

Therefore, Ramirez et al. (2012a, b) proposed a Subcategory Assessment Method (SAM) to reduce the variability in the subcategories assessment in S-LCA studies. In the current chapter, a study based on SAM for a bottle of wine produced by a family Italian winery is presented in order to test it.

11.2 SAM

SAM assesses the social profile of the organizations involved in the processes along the product life cycle in relation to the fulfillment of a Basic Requirement (BR; Ramirez et al. 2012a). In SAM, the BR is a reference point/threshold that enables us to understand the significance of the data collected in the inventory phase. The BR for each subcategory is defined according to methodological sheet indicators (UNEP and SETAC 2010), which may be international agreements, national laws, etc.

In order to provide an objective assessment, SAM is based on a four-level scale for each subcategory which enables the organization to be assigned to Levels A, B, C or D according to its practices. These levels are necessary in order to support the practitioner in applying the method with specific and clear information on how to assess the organizational practices in a more standardized manner for all evaluations.

Levels A and B correspond to an organization that achieves the BR. Level A means that the organization shows a proactive behavior compared to the BR. In SAM, A is assigned to the organization which promotes good practices along all the supply chain related to each subcategory. Level B means that the organization fulfills the BR. Levels C and D identifies an organization which does not meet the BR. The difference between C and D depends on the social conditions of the context (either country or sector). SAM uses the context data available at the Green Delta TC Social LCA database 2011. When this data is not available, the context is defined on information from the organization itself.

11.3 Stakeholder Worker

Eight subcategories refer to "worker" as a stakeholder group: *freedom of association and collective bargaining, child labor, fair salary, working hours, forced labor, equal opportunities/discrimination, health and safety, social benefits/social security* (UNEP and SETAC 2009).

The BR is based on Conventions from International Labour Organization (ILO; ILOLEX 2012).

SAM levels are organized related to each subcategory as in Table 11.1 For example, regarding the subcategory "Freedom of association and collective bargaining subcategory", the BR is the evidence that employees of the organization are associated with a union (at least one), as under ILO Convention no. 87 (ILOLEX 2012). Therefore, if this requirement is met during the evaluation process, the organization is assessed as B. For this subcategory, the difference between C or D assessment depends on the Worker Rights Score (WRS) of the country where the organization is located. The WRS is an indicator used by CIRI Human Rights dataset (CIRI 2012) and refers to the right of collective bargaining, the right to minimum working conditions, freedom of assembly and association, including the rights of citizens to assemble freely and associate with other people in political parties, trade unions, cultural organizations, or other groups. This index ranges from 0 to 4, with the lowest score considered the worst condition, and the highest score, the best one. In SAM, the countries were split into two levels: those scored from 0 to 2 (bad WRS) and those scoring from 3 to 4 (good WRS). Therefore, if an organization does not meet the BR in a country with a good WRS, it is assessed as D. Otherwise, it is assessed as C.

Table 11.1 SAM for stakeholder worker

Subcategory	Level B (basic requirement)	Level C	Level D
Freedom of association and collective bargaining	In the organization there is evidence of workers being associated in the worker's union	Worker rights score of the country where the organization is located is between [0; 2.9]	Worker rights score of the country where the organization is located is between [3; 4]
Child labor	For developed countries Minimum Age for Admission to Employment is 15 years of age. Developing and least developed countries a Minimum Age for Admission to Employment is 14 years of age	There is evidence of child labor in the organization, but it is not defined as the "worst forms of child labor"; and the child attends school	There is evidence of child labor defined in the "worst forms of child labor" or the child does not attend school
Forced labor	The organization has no evidence of use of forced labor	The organization has evidence of forced labor as well as the country where the organization is located	The organization has evidence of forced labor and the country where the organization is located has no presence of forced labor
Fair salary	The worker's wage equal to the minimum wage on the country/sector where the organization is located	The country has "GDP on a purchasing power parity basis, divided by population" smaller than the "living wage x purchasing power parity conversion factor"	The country has "GDP on a purchasing power parity basis, divided by population" higher than the "living wage x purchasing power parity conversion factor"
Working hours	The average weekly hours worked equal to 48 and 8 h on the day or limit is according to the national law	The average weekly hours worked is higher than 48 and lower than the average weekly hours worked for the sector/country	The average weekly hours worked is higher than 48 and higher than the average weekly hours worked for the sector/country
Equal opportunities/discrimination	The organization has a management system, policy or actions to avoid the discrimination and promotes the equal opportunities for the workers	The organization has evidence of discrimination and the country where the organization is located has a GEI score lower than 50	The organization has evidence of discrimination and the country where the organization is located has a GEI score equal to or higher than 50

Table 11.1 (continued)

Subcategory	Level B (basic requirement)	Level C	Level D
Health and safety	The organization meets the national law related to health and safety	Occupational accidents of the organization (rates of injuries and occupational injuries) are smaller than the Occupational accidents of the country/sector (rates of injuries and occupational fatal injuries) where the organization is located	Occupational accidents of the organization (rates of injuries and occupational fatal injuries) are equal or higher than the Occupational accidents of the country/sector (rates of injuries and occupational fatal injuries) where the organization is located
Social benefits/social security	Social benefits (more than two): Retirement, Disability, Dependents, Survivors benefits, Medical insurance, Dental insurance, Paramedical insurance including preventive medicine, Medicine insurance, Wage insurance, Paid maternity and paternity leave, Paid sick leave, Education and training	The organization fulfils at least two items of the Social benefits/social security basic requirement	The organization does not fulfill any item of the Social benefits/social security basic requirement or the organization has workers without an employment contract

11.4 Stakeholder Consumer

Five subcategories refer to "consumer" as a stakeholder group: *health and safety, feedback mechanism, consumer privacy, transparency, end of life responsibility* (UNEP and SETAC 2009). In SAM, only the consumers of the final product are taken into account.

The BR, in this case, is based on International Agreements, such as ISO 26000 (ISO 2010), Global Reporting Initiative (GRI 2006) and Consumer Protection Act (DTI 2011).

SAM levels are organized related to each subcategory (Table 11.2). For instance, for the *consumer privacy* subcategory, the BR is that the organization protects the consumer's right to privacy in its policy, according to Universal Declaration of Human Rights, Article 12 (UDHR 2007), and Consumer Protection Act (DTI 2011). If the organization does not meet the BR, it can be assessed as C or D, according to the Privacy International Ranking score of the country where the organization is located (Privacy International 2007). In this method, the countries were split in two levels: those that are scored from 1.1 to 3 (I) and those that are scored from 3.1 to

Table 11.2 SAM for stakeholder consumer

Subcategory	Level B (basic requirement)	Level C	Level D
Consumer privacy	The organization protects the consumer's right to privacy through a privacy policy	The organization does not meet the basic requirement; and the country where the organization is located has Privacy International Ranking score [1.1; 3]	The organization does not meet the basic requirement; and the country where the organization is located has Privacy International Ranking score [3.1; 5]
Health and safety	The organization is in compliance with national law regarding consumer product health and safety standards	The organization does not meet the basic requirement, but has no proven cases that violate the consumer health and safety	The organization does not meet the basic requirement; and has presence of proven cases that violate the consumer health and safety
Feedback mechanism	The organization has customer feedback mechanism and practices related to customer satisfaction. It provides all the following practices: suggestion box on help desk, customer satisfaction surveys, complaint service and/or section on the website	The organization does not meet the basic requirement and there is no record of consumers' complaints regarding the lack of Feedback mechanism	The organization does not meet the basic requirement and there is record of consumers' complaints regarding the lack of Feedback mechanism
Transparency	The organization has a report that communicates social responsibility	The organization does not meet the basic requirement, but it has ways of showing their consumer technologies, good practices and management conduct. For example: through events or web site information	The organization does not have a report that communicates the social responsibility nor ways of showing their consumers technologies, good practices and management conduction
End of life responsibility	There are internal management systems that provide clear information to consumers on end-of-life options	The organization does not meet the basic requirement; but the end of life product can be considered recycled by the municipal selective collection	The organization does not meet the basic requirement; and the end of life product is not recycled by the municipal selective collection. For example: batteries

5 (II). Countries assessed as I are those with worse conditions with regard to the national privacy ranking. Those organizations located in countries in I are assessed as C because they are in accordance with the country context. On the other hand, the organizations that are in a country where the context is adequate, receive a D.

11.5 Case Study in the Wine Sector

The wine sector represents one of the most important pillars of the agro-food sector of several countries. In Italy, the wine sector is distinguished by low level production process standardization in vineyards and wine-producing facilities. Moreover, this sector has few large companies as most producers are medium or small-sized and their governance models are often family-related (Petti et al. 2010).

In 2008, world wine production was 27.27 Mt, mainly concentrated in Europe (64 %; Petti et al. 2010).

The trend is towards reduced quantity but higher quality (FAO 2008) and in Italy the production trend is moving towards high quality wines as well.

The case study was developed in partnership with a wine producer located in the hills of the Abruzzo region. It is a small winery managed by the family and the vineyards are worked by hand where possible, allowing for greater care of the quality of the vines during all work phases.

The goal of this study is to analyze the stakeholder consumer and worker of the grape and wine production phases using SAM. Therefore, it is a cradle-to-gate study. The unit of analysis considered is the production of 75 mL of wine ("novello" wine, produced and consumed within a year). The site data collection was performed by interviewing managers and technicians of the wine company and representatives of consumer organizations. A questionnaire, based on the methodological sheets (UNEP and SETAC 2010), was prepared and sent to the wine company and the consumer organizations to be filled out.

Generic data (data from the country and wine sector) was obtained from consumer organizations, national statistics and national laws regarding consumer and workers' rights. The resulting information of the data assessment using SAM is summarized in Table 11.3.

The assessment highlights that for the stakeholder consumer the organization for most of the subcategories did not reach the basic requirement. For example, as regarding the subcategory health and safety, the organization is not in total compliance with the national law concerning consumer product health and safety standards, as a test required by EC 123/5 (EU 2005) is not performed for each crop of grapes. On the contrary, all other requirements for this subcategory are fulfilled. Therefore, according to Table 11.2, the organization is rated as C (the organization does not fulfill the BR). Nevertheless, in some cases, simple changes could improve its performance. For instance, in the subcategory feedback mechanism, the company need only to implement measures which enable the consumer to make complaints, such as providing a suggestion box on the help desk or a customer care section on the

Table 11.3 Wine production assessment using SAM

Stakeholder	Subcategory	Grape production	Wine production
Consumer	Health and safety	–	C
	Feedback mechanism	–	C
	Consumer privacy	–	B
	Transparency	–	C
	End of life responsibility	–	C
Worker	Freedom of association and collective bargaining	C	C
	Child labor	B	B
	Forced labor	B	B
	Fair salary	B	B
	Working hours	B	B
	Equal opportunities/ discrimination	–	–
	Health and safety	B	B
	Social benefits/social security	B	B

B means to fulfill the BR, *C* means not to fulfill the BR

website. This would allow the organization's evaluation to rise from position C to B. The subcategories transparency and end of life responsibility can be interpreted in the same way.

As regards the stakeholder worker in both grape and wine production processes, the organization fulfilled the BR for almost all subcategories, and was thus rated B. The only exception was the subcategory Freedom of association and collective bargaining, which was rated as C as none of the workers were members of the workers' union.

The assessment is standardized around the result B. This can be attributed to the fact that it was not possible to carry out the triangulation of data with the workers. Assessment is lacking for the subcategory equal opportunities/discrimination due to the fact that the questionnaires regarding this subcategory were not completed.

11.6 Conclusions

This first SAM application on the wine sector clearly shows in which phases and subcategories the basic requirements were achieved and in which they were not.

The case study also underlines that it was possible to collect the data and to evaluate the organization regarding the stakeholder consumer and worker using SAM. This enables an understanding of the strengths and weaknesses of each phase of the product. Moreover, it demonstrates how SAM transforms qualitative data to usable information to assist in a product overview, which could help in decision making.

S-LCA is even more time and work demanding than (environmental) LCA, because the issues are too specific and change from company to company, sector to sector and region to region.

The method may be implemented for the entire life cycle of a product, which necessarily involves both more processes and more organizations. The future development of SAM will include the remaining subcategories and adapt the basic requirement for each of them as well as considering the organization's conduct within the context in which it operates.

However, this case study highlights that not all the specific situations (regionalization elements, background and organization size) can be easily taken into account with this methodology, especially when considering family agriculture. This is due to the fact that SAM is based on the assessment of organizational practices in relation to international agreements and/or national laws; while this contributes to the objectivity of the method, it is also a limitation. This was found to be the case when assessing the stakeholder consumer, as most of the basic requirement was not fulfilled because of the way a family organization may manage its business.

References

Ciroth A, Franze J (2009) Social life cycle assessment of roses—a comparison of cut roses from Ecuador and the Netherlands. Presentation, Life Cycle Assessment Conference Boston IX, 29 September–2 October 2009

Ciroth A, Franze J (2011) LCA of an ecolabeled notebook—consideration of social and environmental impacts along the entire life cycle. GreenDeltaTC, Sustainable Development, Berlin

CIRI (2012) The CIRI human rights data project. http://ciri.binghamton.edu/index.asp. Accessed 10 Jan 2012

DTI—Department of Trade and Industry (2011) Consumer protection act. http://www.issafrica.org/crimehub/uploads/Consumer_protection_PocketGuide.pdf. Accessed 20 Nov 2011

Dreyer LC, Hauschild M, Schierbeck J (2006) A framework for social life cycle impact assessment. Int J Life Cycle Assess 11(2):88–97

EU—European Union (2005) Commission regulation reg. n° 123/05. http://eurlex.europa.eu/LexUriServ/LexUriServ.do?uri=OJ:L:2005:025:0003:0005:EN:PDF. Accessed 20 Nov 2011

FAO (2008) http://faostat.fao.org. Accessed 10 Oct 2009

GRI—Global Reporting Initiative (2006) Sustainability reporting guidelines G3.1—reference sheet. http://www.globalreporting.org/NR/rdonlyres/ED9E9B36-AB54-4DE1-BFF2-5F735235CA44/0/G3_GuidelinesENU.pdf. Accessed 10 Oct 2011

Hunkeler D (2006) Societal LCA methodology and case study. Int J Life Cycle Assess 11(6):371–382

ILOLEX (2012) Database of International Labour Standards. http://www.ilo.org/ilolex/english/convdisp1.htm. Accessed 10 Jan 2012

ISO (2010) ISO 26000: Guidance on social responsibility. International Organization for Standardization, Geneva

Norris G (2006) Social impacts in product life cycles: towards life cycle attribute assessment. Int J Life Cycle Assess 11(1):97–104

Petti L, Ardente F, Bosco S, De Camillis C, Masotti P, Pattara C, Raggi A, Tassielli G (2010) Stato dell'arte della Life Cycle Assessment (LCA) nel comparto vitivinicolo. In: Francesca Cappellaro e Simona Scalbi (eds), La metodologia LCA: approccio proattivo per le tecnologie

ambientali—Casi studio ed esperienze applicative. Proceedings Scientific Conference of Rete Italiana LCA, Padova, 22 April 2010, ENEA, Rome
Privacy International (2007) Privacy International. https://www.privacyinternational.org/survey/rankings2007/phrcomp_sort.pdf. Accessed 18 Nov 2011
Ramirez PKS, Petti L, Brones F, Ugaya CML (2012a) Método de Avaliação de Subcategoria em ACV Social: Aplicação para Trabalhadores no Sabonete de Cacau da Natura. In: Proceedings of "III Congresso Brasileiro de Gestão do Ciclo de Vida", Maringá, Brasil
Ramirez PKS, Petti L, Ugaya CML (2012b) Subcategory assessment method: stakeholder workers. In: Corson MS, van der Werf HMG (eds) Proceedings of the 8th International Conference on Life Cycle Assessment in the Agri-Food Sector (LCA Food 2012), 1–4 October 2012, Saint Malo, France, INRA, Rennes, France
UDHR—Universal Declaration of Human Rights (2007) The universal declaration of human rights. http://www.un.org/events/humanrights/2007/udhr.shtml. Accessed 18 Nov 2011
UNEP, SETAC (2009) Guidelines for social life cycle assessment of products, United Nations Environment Program, Paris, SETAC Life Cycle Initiative, United Nations Environment Programme
UNEP, SETAC (2010) Methodological sheets of sub-categories of impact for a Social LCA. http://lcinitiative.unep.fr. Accessed 18 Nov 2011
Weidema B (2005) ISO 14044 also applies to social LCA. Int J Life Cycle Assess 10(6):381
Weidema B (2006) The integration of economic and social aspects in life cycle impact assessment. Int J Life Cycle Assess 11(1):89–96

Chapter 12
Human and Environmental Impact Assessment for a Soybean Biodiesel Production Process Through the Integration of LCA and RA

Francesco Spina, Giuseppe Ioppolo, Roberta Salomone, Jan C. J. Bart and Maria Francesca Milazzo

Abstract Biodiesel is one of the most important global renewable fuels produced by vegetable biomass. It is the main green fuel produced and utilized in Europe (in this context Italy is the 4th producer country) and the result of national biofuel mandates. The EU Renewable Energy Directive imposes to reach a minimum threshold of 10% diesel blending with biodiesel by 2020 in order to reduce global warming. Biodiesel has many positive aspects, it is a renewable, non-ecotoxic and obtained from vegetable oils and animal fats with a favorable balance in terms of GWP (Global Warming Potential) and EROI (Energy Return On Investment), but its production has also some critical aspects mainly related to its impacts on the environment and health. The aim of this chapter is to demonstrate how an approach for the assessment of these impacts, by means of the integration of LCA (Life Cycle Assessment) and RA (Risk Assessment) methodologies, helps in evaluating the acceptability of the process.

Keywords Biofuel · Life Cycle Assessment · Risk Assessment · Human health · Soybean oil

12.1 Introduction

Biofuels are derived from biomass. There are many kinds of biofuels, such as bioethanol, biodiesel and biomethane. Bioethanol is obtained by alcoholic fermentation of sugars, biodiesel is produced by a chemical process from vegetable oils or animal fats, and biomethane derives from an anaerobic degradation of organic waste. Biofuels are intended to substitute traditional fuels derived from fossil

M. F. Milazzo (✉) · F. Spina · J. C. J. Bart
Department of Electronics Engineering, Chemistry and Industrial Engineering (DIECII), University of Messina, 98166 Messina, Italy
e-mail: mfmilazzo@unime.it

G. Ioppolo · R. Salomone
Department of Economics, Business, Environment and Quantitative Methods, University of Messina, Piazza S. Pugliatti 1, 98122 Messina, Italy

R. Salomone, G. Saija (eds.), *Pathways to Environmental Sustainability*,
DOI 10.1007/978-3-319-03826-1_12, © Springer International Publishing Switzerland 2014

resources for a variety of reasons. Biodiesel has many environmental benefits and, due to the increase of the price of diesel, the price at the pump of biodiesel has almost reached that of conventional diesel (Babcock 2012), causing a growth of the biodiesel production in recent years (Lamers 2011). Disadvantages are due to the handling of hazardous substances during production potentially leading to undesired consequences for both people and the environment, as well as to the use of fertilizers and other agrochemicals during cultivation of the oilcrops.

The effects due to potential release of greenhouse gases to the environment are usually assessed using LCA (Life Cycle Assessment) and RA (Risk Assessment) approaches. The first method is applied to analyze the life cycle of the product, understand the critical points of the production process and improve the performance of the system; the second method analyses human and environmental risks. By integrating the two approaches to study a production process it is possible to exploit the potential of both in order to provide a more complete impact assessment. In particular, there is a need to develop integrated methodologies to assess environmental impacts of innovative materials and emerging technologies. In the literature there are few attempts to integrate LCA/RA (see Sect. 12.2.3), as this integration is not easy due to many aspects that are described in this chapter. After a review of previous studies, the objective of this chapter is to suggest a practical procedure to combine both approaches. Application to soy biodiesel production allows to evaluate the acceptability of the process.

12.2 Methods for the Impact Assessment of Industrial Processes

As mentioned above the objective of this c is to suggest a practical procedure to combine LCA and RA to exploit the potential of both approaches in providing a more complete assessment of the impacts. In this section a brief description of these methods is given, and a review of literature studies related to the most significant approaches for the integration LCA/RA is reported.

12.2.1 Life Cycle Assessment

LCA is a method standardized by ISO 14040 to study the impact of a process by analyzing input and output to the system. LCA is largely a site-generic assessment tool at a global/regional scale, not location or time-specific. It is a tool to estimate environmental burdens and is data intensive.

The method consists of the following four steps:

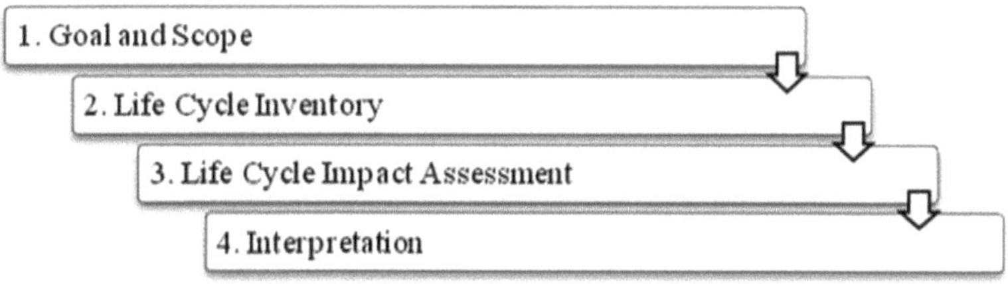

In the first phase the purpose and scope of the LCA study have to be defined. The second step is the collection of data, this must be referred to a functional unit (which is the reference to which all other data are normalized). The third phase is the Impact Assessment (IA), in which a number of impact categories have to be selected, defined, estimated, classified and characterized by aggregating the inventory results in terms of adequate factors of different types of substances within the impact categories. The last step is the Interpretation; it aims to evaluate the results from the inventory analysis or impact assessment and compare them with the goal of the study defined in the first phase (Sonnemann et al. 2004).

12.2.2 Risk Assessment

RA is the process to estimate the likelihood of occurrence of adverse effects to humans and ecological receptors as a result of exposure to hazardous chemicals, physical and/or biological agents (US EPA 1991). It is applied at a local scale. The methodology comprises the following steps:

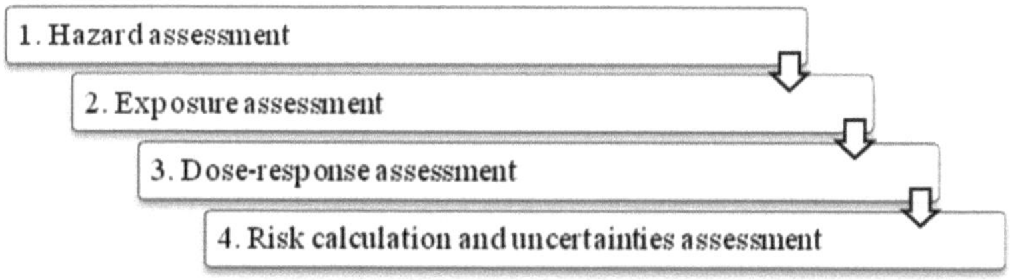

The hazard assessment step identifies the hazards associated with the release or the presence of chemicals in the environment. The goal of the exposure assessment phase is to quantify the intensity, frequency and duration of exposure for people and ecosystems; the exposure is related to the contact receptors/substances and aims at analyzing the effective quantity of contaminant absorbed by the receptor. The third phase, the dose-response assessment, determines the relationship between the exposure and the expected occurrence of the undesired effect. The dose (or intake dose) is the amount of substance received by the receptor during the environmental exposure by inhalation, ingestion or dermal absorption. Finally, the risk calculation phase consists of the integration of all the results of the previous phases to obtain the risk description which, for example, can show the increased probability of developing cancer during a lifetime as a result of the exposure to hazardous chemicals, physical and/or biological agents. Risk calculations must be completed with an uncertainty assessment (see Milazzo and Aven 2011).

12.2.3 Approaches for the Integration of LCA and RA

Due to some limits of both methodologies, the need of integrating LCA and RA has grown to give an holistic approach to quantify environmental impacts. As shown in the previous sections the main characteristics of the methods are different, but by combining them it is possible to overcome some of their limitations and to obtain benefits. LCA provides information to be used for product improvement and minimization of potential impacts (allows comparison between innovative and conventional processes). RA focuses on chemical substances and their effects on the environment and human health and identifies risk due to certain activities in a specific site during a given period. It is not easy to define similarities and differences, but it is clear that: (i) they are used for a different typology of analyses, (ii) their scope and perspectives are different and (iii) the connection between them is not well established.

From an analysis of both methodologies it can be concluded that the connection between them mainly regards the following aspect. The LCIA step of LCA models potential impacts related to the functional unit and uses different impact categories, whereas RA assesses only impacts on humans and ecosystems. These impacts are also estimated by LCA, but the method simplifies the assessment of potential effects because they do not include detailed information, such as the quantification of the exposure (see Sect. 12.2.2).

The literature reports many examples of LCA/RA integration, but in this section we will illustrate only the most significant ones. In this context the work of Flemström et al. (2004) gives a valid support for combining LCA and RA. Approaches for the LCA/RA integration can be classified by means of Venn diagrams, as shown in (Fig. 12.1).

Using approach A of Fig. 12.1, the methods are completely separated and each of them is applied with its specific framework. Approach B shows a partial overlap due to the use of some common data. In approaches C and D, one method is a subset of the other. In particular, case C means that toxicological models from RA are used in LCIA, whereas case D is not used because RA is limited to a specific time and site and cannot be included in LCA (which refers to more general aspects). Finally, using approach E, RA and LCA are considered as complementary tools, which means that a detailed risk assessment is made for each emission identified through LCA.

Once the approach is defined, hazardous substances are involved in a number of different ways during life cycle of a product. They can be the product itself or a by-product, a raw material, an emission due to the production process, a sub-component

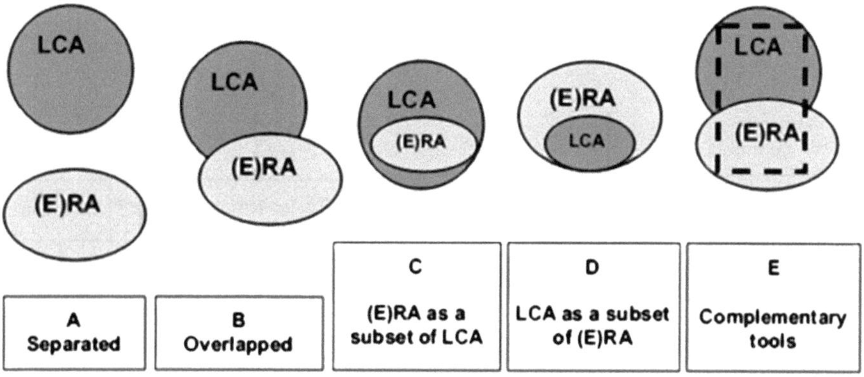

Fig. 12.1 Approaches for combining Life Cycle Assessment and Risk Assessment. (Flemström et al. 2004)

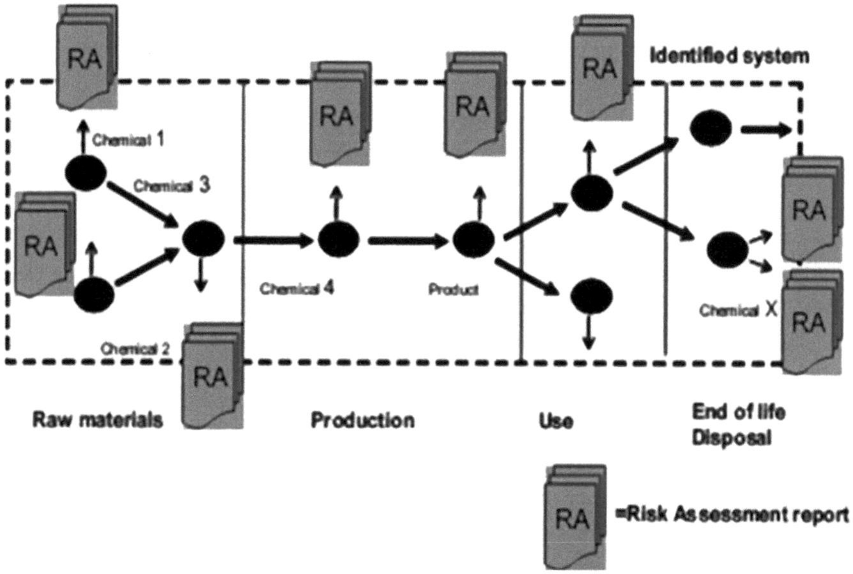

Fig. 12.2 Integration of Life Cycle Assessment and Risk Assessment as proposed by Flemström et al. (2004)

of an inflow or outflow, waste, etc. Flemström et al. (2004) proposed to perform a detailed risk assessment for each emission identified in the life cycle chain of a product or service (Fig. 12.2).

The method proposed by Nishioka et al. (2005) aims to incorporate regional variability in emissions and exposure in LCA, using relationships derived from risk assessment methods. They calculated intake fractions and applied these values to the supply chain emissions. Then they estimated population exposure only for the pollutants passing a screening risk level. The same authors (Nishioka et al. 2005a)

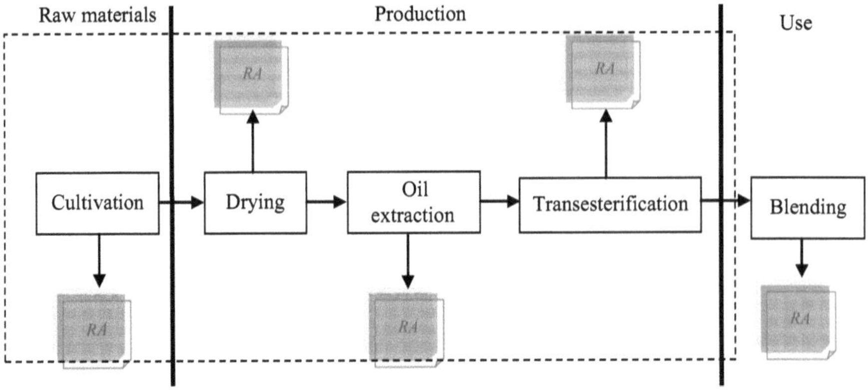

Fig. 12.3 Biodiesel production process: boundaries of the system

applied this methodology to a case-study and demonstrated how their approach helps decision-makers in evaluating the costs and benefits according to the public health and environmental policies.

Udo de Haes et al. (2006) proposed a three levels approach to describe the linkages between RA and LCA. At the first level few differences exist because both tools use essentially the same fate and effect models, including their coefficients and data; in the second level the application of the analysis has many differences due to the diversity in the scope, types of impacts, use of characterization factors, spatial and temporal detail aggregation of effects and the functional units as a basis of the assessment; at the third level the aims of the tools are complementary.

Matthews et al. (2002) reported that the output of a life cycle inventory analysis is the quantity of various contaminants going into the air, water, and land-fills during each stage of the life cycle. Thus, the challenge is to infer how to proceed from environmental discharges to the quantification of associated risks. The authors considered that developing robust methods to translate the inventory of environmental discharges into an inventory of environmental impacts is a challenge to risk analysis.

12.3 The Analysis of the Impact of a Biodiesel Production Process

A practical procedure to combine LCA and RA for the study of the biodiesel production process derives from the work of Flemström et al. (see Fig. 12.2). We analyzed a real case-study related to the biodiesel production from soybean and, to define the impact of the process the boundaries of the system were delimited as reported in Fig. 12.3. The raw material for the process is the soybean from which the oil is extracted, its production requires the use of diesel fuel, seeds, fertilizers and pesticides, whose transports are not included in the study.

Table 12.1 LCI for biodiesel production

Chemical	Raw materials (g per kg of soybean)	Production (g per kg of biodiesel)	
	Cultivation	Drying, extraction, refining	Trans-esterification
Air emissions			
Carbon dioxide (CO_2 fossil)	182.87	370.41	
Carbon dioxide (CO_2 biomass)	1832.98	0	
Methane (CH_4)	0.181	0.874	
Nitrous oxide (N_2O)	0.008	0.003	
Carbon monoxide (CO)	0.873	0.123	
Hydrocarbons (except CH_4)[a]	0.784	2.015	
Benzene[a]	8.38 E-06	0	
Formaldehyde[a]	1.13 E-04	4.17 E-12	
Particulates (PM10)	0.087	0.006	
Particulates (unspecified)	0.099	0.438	
Sulphur oxides (SO_x as SO_2)[a]	0.599	2.454	
Nitrogen oxides (NO_x as NO_2)[a]	1.284	0.784	
Hydrogen chloride (HCl)[a]	0.002	0.017	
Hydrogen fluoride (HF)	7.83 E-05	0.002	
Ammonia (NH_3)[a]	0.469	4.62 E-05	
Water emissions			
BOD5	0.034	4.16 E-04	
COD	0.240	0.003	
Methanol (CH_3OH)[a]	–	0.8572	
Metals (K and Na)	0.001	0.002	
Ammonia (NH_4^+, NH_3, as N)[a]	0.004	8.26 E-05	
Nitrates (NO_3^-)	1.71 E-05	2.21 E-05	

[a] Substances included in RA

The analysis ends with the use of biodiesel by blending with the conventional petrodiesel (neglecting the use and end-of-life of the blended product which is combustion in a car). The action is located within an oil refining plant producing the conventional diesel. This permits to exclude transport of biodiesel from the analysis as it is used in situ.

To proceed with the LCA/RA integration, emissions related to the various stages of production are necessary. Our study has been limited to the analysis of emissions in air and water. Table 12.1 shows the life cycle inventory, taken from Sheehan et al. (1998). Using this data and specific transport models for the carcinogenic/toxicological and eco-toxicological contaminants, RA can be applied related to substances that are harmful for human health, which are highlighted in gray in the table. The selection of these contaminants has been carried out by reference to the annexes of the current legislation (see Italian Decree Law no. 152/06). The selected substances are those that, following the release into the environment, can cause damage to human health by inhalation, ingestion or dermal contact (exposure routes).

As reported by Sheehan et al. (1998), the crushing operation is directly responsible for 85 % of the hydrocarbons emissions. Virtually all these emissions are attributable to the loss of hexane through vents and leaks.

In order to integrate RA/LCA, it is possible to use human toxicity potentials (HTPs) and the release inventory. Thus correlations for the estimation of impacts are derived from risk assessment methodology. HTPs proposed by Guinée and Heijungs (1993) present an hazard evaluation based on toxic potency of substances (inverse of the allowable daily intake) and potential doses. The exposure is calculated using multimedia environmental fate models, and therefore HTP is significantly closer to an actual risk assessment than other approaches. The individual lifetime risk is defined as follows (i is the intake route, k is a compartment different from the exposure medium n, p is exposure pathways):

$$\mathrm{H}(C_{cn}) = \sum_{\substack{i \text{ (intake} \\ \text{routes)}}} \sum_{\substack{k \text{ (environmental} \\ \text{compartment)}}} \sum_{\substack{p \text{ (exposure} \\ \text{pathway)}}} \left(Q_{ci} \cdot \frac{ADD_{cip}}{C_{ck}} \cdot \Phi\left(C_{cn}(0) \rightarrow C_{ck}\right) \right) \quad (12.1)$$

where Q_{ci} is the dose-response function; ADD_{cip} is the average daily potential dose (over a specified averaging time) from an exposure medium n by route i (inhalation, ingestion, dermal contact) attributable to the environmental compartment k; $C_{cn}(0)$ is the contaminant concentration in the emission at initial time (zero); C_{ck} is the contaminant concentration in the compartment k at time t; Φ is the environmental fate function (multimedia dispersion function), which converts the actual contaminant concentration $C_{cn}(0)$ measured (mg/kg) into future contaminant concentration C_{ck} at a time t (exposure time ED).

HTP is the individual lifetime risk (H_{cn}) resulting from a unitary emission (1 kg/d) of chemical c into the compartment specific n normalized by the individual lifetime risk resulting from the unitary emission of a reference substance (benzene for carcinogenic effects and toluene for non-carcinogenic effects) to a reference compartment (such as air):

$$\mathrm{HTP}_{cn} = \frac{\mathrm{H}_{cn}(C_{cn}(0) = 1)}{\mathrm{H}_{reference,air}(C_{reference,air}(0) = 1)} \quad (12.2)$$

The final aggregate HTP is obtained as:

$$\mathrm{HTP} = \sum_{c \text{ (chemical)}} \sum_{\substack{n \text{ (release} \\ \text{compartment)}}} \mathrm{HTP}_{cn} \cdot C_{cn}(0) \quad (12.3)$$

12.4 Results and Discussion

Using data of Table 12.1 and Eqs. (12.2) and (12.3), as suggested by Hertwich et al. (2001), the impact of substances included in RA on human health has been calculated related to our case-study and the results are shown in (Table 12.2).

Table 12.2 Impact of substances on the human health (carcinogenic and non-carcinogenic effects)

Chemical	Raw materials		Production	
	HTP carcinogenic	HTP non carcinogenic	HTP carcinogenic	HTP non carcinogenic
Air emissions				
Hexane	–	–	–	3.44 E-01
Benzene	1.55 E-06	2.64 E-05	–	–
Formaldehyde	6.28 E-08	1.00 E-04	6.95 E-19	1.11 E-09
Sulphur oxides (SO_x as SO_2)	–	8.21 E-05	–	1.01 E-04
Nitrogen oxides (NO_x as NO_2)	–	2.38 E-03	–	4.35 E-04
Hydrogen chloride (HCl)	–	1.59 E-04	–	4.06 E-04
Ammonia (NH_3)	–	2.78 E-01	–	3.55 E-04
Water emissions				
Methanol	–	–	–	1.38 E-03
Ammonia (NH_4^+, NH_3, as N)	–	4.37 E-05	–	2.71 E-07

As shown in Table 12.2, the highest risk values are associated with emissions of nitrogen oxides and ammonia into the air, in particular, during the cultivation phase, and also to the oil extraction (hexane) and the (methanol) transesterification processes. Ammonia and NO_x derive from agrochemicals used for the soybean cultivation, in particular fertilizers or organic manure. Ammonia is detected also in water emissions.

Results of this study give a more complete evaluation of the acceptability of the biodiesel process than that obtained through the mere application of LCA. Concerning this bio-fuel, literature shows the following: GHG emission reduction of US soy biodiesel is about 56 % compared to the fossil reference (OMNI TECH/United Soybean Board 2009)

In addition, according to data of this study, the process is acceptable both from a human health and ecological point of view. A more complete analysis is underway comparing our results with those related to production of diesel.

12.5 Conclusions

The evaluation of the potential environmental and human health impacts through the integration RA/LCA has allowed to extend considerations about the acceptability of the biodiesel production process from soybean oil compared to those derived from the application of the individual techniques. Despite many LCA studies, there are no reports in the literature analyzing also the aspects related to risk assessment (RA), which are crucial for understanding the criticalities of the process. A comprehensive study of the impacts cannot ignore them, as shown in this chapter.

Concerning our study, it has been concluded that in the overall assessment of the process, a significant impact on health (toxicological, non-carcinogenic) is associated with emissions of NO_x and ammonia related to the cultivation step.

References

Babcock S (2012) Clean cities alternative fuel price report. U.S. Department Of Energy. http://www.afdc.energy.gov/uploads/publication/afpr_jul_12.pdf. Accessed 15 Sept 2012

Flemström K, Carlson R, Erixon M (2004) Relationships between life cycle assessment and risk assessment—potentials and obstacles. Report 5379, Industrial Environmental Informatics (IMI), Chalmers University of Technology

Guinée J, Heijungs R (1993) A proposal for the classification of toxic substances within the framework of life cycle assessment of products. Chemosphere 26:1925–1944

Hertwich G, Mateles SF, Pease WS, McKone TE (2001) Human toxicity potentials for life-cycle assessment and toxics release inventory risk screening. Environ Toxicol Chem 20(4):928–936

Lamers P (2011) International biodiesel markets. Berlin: UFOP-Schriften. http://www.ecofys.com/files/files/ecofys_ufop_2012_internationalbiodieselmarkets.pdf. Accessed 15 Sept 2012

Matthews S, Lave L, MacLean H (2002) Life cycle impact assessment: a challenge for risk analysts. Risk Anal 22(5):853–860

Milazzo MF, Aven T (2011) An environmental risk assessment of a contaminated site based on extended uncertainty analyses. Proceedings of the European Safety Reliability Conference ESREL 2011 Troyes, ebook, pp 2121–2129

Nishioka Y, Levy IJ, Norris GA, Bennett DH, Spengle JD (2005) A risk-based approach to health impact assessment for input-output analysis. Part 1: methodology. Int J LCA 10:193–199

Nishioka Y, Levy IJ, Norris GA, Bennett DH, Spengle JD (2005a). A risk-based approach to health impact assessment for input-output analysis Part 2: case study of insulation. Int J LCA 10:255–262

OMNI TECH Intl. and Four Elements Consulting LLC (to United Soybean Board) (2009) Applying Recent US Soybean Data to the EU Renewable Energy Sources Directive, Midland, OH

Sheehan J, Camobreco V, Duffield J, Graboski M, Shapouri H (1998) Life cycle inventory of biodiesel and petroleum diesel for use in an urban bus, National Renewable Energy Laboratory, Golden, CO, NREL/SR-580-24089

Sonnemann G, Castells F, Schuhmacher M (2004) Relationships between life cycle assessment and risk assessment. Swed Environ Prot Agency

Udo de Haes HA, Wegener SA, Heijungs R (2006) Similarities, differences and synergisms between HERA and LCA—an analysis at three levels. Hum Ecol Risk Assess 12:431–449

US EPA (1991) Risk assessment guidance for Superfund: volume 1—human health evaluation manual (Part B, Development of risk-based preliminary remediation goals). EPA/540/R-92/003. United States Environmental Protection Agency, Washington, DC

Zhiyuan H, Piqiang T, Xiaoyu Y, Diming L (2008) Life cycle energy, environment and economic assessment of soybean-based biodiesel as an alternative automotive fuel in China. Energy 33:1654–1658

Chapter 13
Energy and Environmental Assessment of Retrofit Actions on a Residential Building

Maurizio Cellura, Mario Fontana, Sonia Longo, Daniele Milone and Marina Mistretta

Abstract Energy and environmental performances of buildings strictly depend on many factors related to the choice of construction materials, technical equipment, installation and use. In the following chapter a set of retrofit actions to improve the thermal performance of an existing conventional building is presented. The energy and environmental assessment of these actions is carried out following a life cycle approach. The embodied energy and the environmental impacts arisen from the production, transportation and installation phases of the required materials and components are calculated. Further, energy saving and environmental benefits and drawbacks concerning the assessed retrofit actions are accounted for.

Keywords Life cycle assessment · Embodied energy · Environmental impacts · Retrofit · Mediterranean buildings

13.1 Introduction

In EU, the most of Governments adopted new policy tools, starting from the European Commission's Directive on the Energy Performance of Buildings, which introduced environmental performances as the most relevant driving force for energy saving in buildings (climate change, resource depletion, toxicity, etc.) (Directive 2010). Worth of note is also the Directive 2009/125/EC which established a framework to set eco-design requirements for energy related products and to define a general scheme for the integration of environmental issues in the product design (Directive 2009).

The significance of energy consumption and greenhouse gas emissions arisen from building operation is well understood. However, due to the growing awareness that buildings require energy over their lifespan, an exhaustive assessment of

M. Mistretta (✉)
Dipartimento Patrimonio, Architettura, Urbanistica,
University of Reggio Calabria, Reggio Calabria, Italy
e-mail: marina.mistretta@unirc.it

M. Cellura · M. Fontana · S. Longo · D. Milone
Dipartimento di Energia, Ingegneria dell'Informazione e Modelli Matematici,
University of Palermo, Palermo, Italy

R. Salomone, G. Saija (eds.), *Pathways to Environmental Sustainability*,
DOI 10.1007/978-3-319-03826-1_13,

the environmental impacts may not leave out of consideration energy consumption, exploitation of natural resources and pollutant emissions in a life cycle perspective (Sartoni and Hestnes 2007; Suzuki et al. 1998). The improvement of the building performances during operation must still be the primary goal of the design step, but the lower the energy demand the higher is the embodied energy. Measures to reduce the operating energy demand and to increase the energy efficiency of technical equipment lead to an increase in embodied energy of buildings, essentially due to the energy used for the production of the involved materials and technical installations (Dixit et al. 2010). Thus, the relevance of the life cycle approach is apparent to perform a reliable and complete building energy and environmental assessment (Gustavsson et al. 2010; Blengini and Di Carlo 2010).

An overall judgment on building sustainability should encompass all the life cycle phases and should be based on the internationally recognized Life Cycle Assessment (LCA). It is a suitable methodology to estimate the eco-profile of a building, including all the phases in which built structures and facilities are procured and erected, operation, maintenance, renovation, dismantling and waste management at the building end-of-life (Hernandez et al. 2011). The above considerations should be applied to building refurbishment.

Housing renovation should reduce the environmental impacts (e.g., energy and resource consumption, emission of air and water pollutants, waste generation, and noise), increase the indoor comfort, and improve the architectural appearance of the building facades (Ardente et al. 2011). LCA supports designers to select, among different retrofit options, the most effective actions addressed to save energy and to minimize environmental impacts along the whole building life cycle.

Starting from the results of a detailed LCA study of a conventional building (Ardente et al. 2008), in this chapter a set of retrofit actions on the building envelope are proposed to improve the thermal performances of the existing building. The energy and environmental assessment is performed following a life cycle approach and points out the embodied energy and the environmental impacts arisen from the production, transportation and installation phases of the required materials and components. Thus, a balance between energy and environmental benefits and drawbacks concerning the assessed retrofit actions is carried out. Then, it is possible to understand whether the achieved benefits can be confirmed from a life cycle perspective or were exceeded by the environmental burdens of the actions.

13.2 Description of the Existing Building and LCA Main Results

The assessed building is a Mediterranean single-family house, located in Palermo in Southern Italy, 270 m above sea level, and is currently used by three occupants. It was built on one level with a heated area of 110 m^2. The local climate involves hot and wet summers, which affect significantly the energy requirement for the building cooling. The frame structure is made of reinforced concrete with masonry block

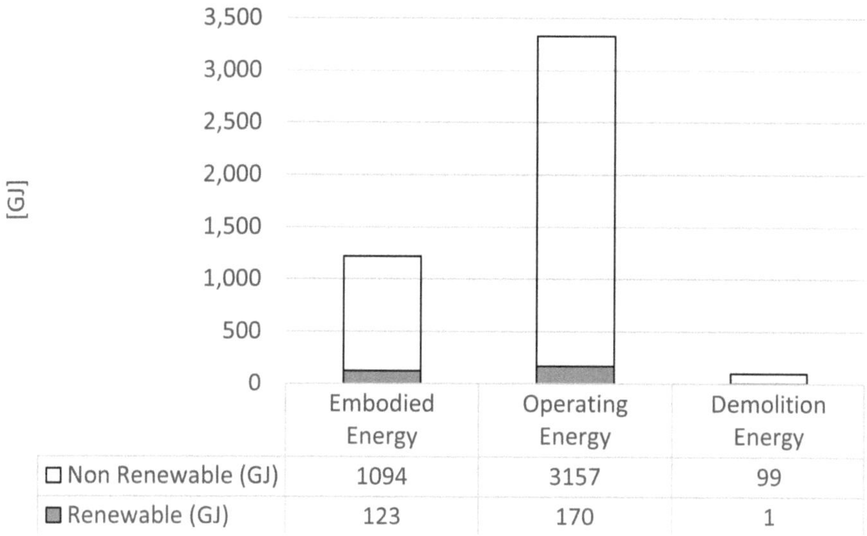

Fig. 13.1 Embodied energy, operating energy and demolition energy in the existing building

Table 13.1 Ecoprofile of the existing building

Indicators	Unit	
CED	GJ	4,645
GWP	kg CO_{2eq}	324,270
ODP	kg CFC_{11eq}	0.05
AP	kg SO_{2eq}	1,193
NP	kg $PO_4^{3-}{}_{eq}$	270
POCP	kg C_2H_{4eq}	378

walls. The external walls include bricks 20 cm thickness with a 9 cm cavity filled with foam vermiculite. The floors are 20 cm thickness, including perforated bricks and prefabricated reinforced concrete rafters. Roof has a wooden structure with composite materials and clay roof tiles cover. The basement lays on a reinforced concrete structure and a layer of cave crushed stones. Table 13.1 shows the eco-profile of the assessed building. The LCA methodology is applied to the building in compliance with the international standards of the ISO 14040 series, including the steps of material and component production, building erection, operation, maintenance and end-of-life (UNI EN ISO 2006). Figure 13.1 illustrates the contribution to the building Cumulative Energy Demand (CED) of the embodied energy, the operating energy and the demolition energy. Embodied energy is estimated as the primary energy used in the raw material acquisition, in the production of materials and technical components, and in the transports to the building site. It also includes the primary energy embodied in the material used in the maintenance phase. Embodied energy represents the 26% of the CED (1,217 GJ). Demolition energy, that is the primary energy consumption in the dismantling and construction waste disposal at end-of-life, accounts only for 2% (about 101 GJ). It is apparent that the operation step involves the highest contribution, accounting for 72% on CED (3,327 GJ). Foreseeing a 50 year lifespan for the building, the operating energy is estimated as the primary energy consumption for specific uses–heating, cooling, Domestic Hot Water (DHW), lighting, electric appliances, and cooking. The monitoring step of the building operation shows that heating, DHW and cooking accounted for about 40%, while electricity for lighting, appliances and cooling total to about 60%. Heating, DHW and cooking are supplied by a Liquid Petroleum Gas boiler.

Heating system is equipped with steel radiant radiators and insulated steel pipes, while summer air cooling is provided with electrical heat pumps installed into main rooms.

The energy requirement for winter heating arisen from the thermal simulation reveals that the thermal performances of the existing building shell are not suitable to comply with the requirements of the current Italian law (Italian Decree 2006). For these reasons the authors propose a set of retrofit actions to increase the thermal transmittance of the building envelope, so that the energy requirement for heating would decrease and keep within the legislative limits.

However, when improving the energy performance of a building, extra materials and components are required, resulting in higher embodied energy. Then, the authors apply LCA methodology in order to assess the overall impact of the proposed retrofit measures over the building lifespan.

13.3 Case Study: Retrofit Actions on a Mediterranean Single-Family Building

13.3.1 Goal and Scope Definition

Following a life cycle approach, the main goals of the study are:

- to assess the energy and environmental impacts of the retrofit actions;
- to analyze the relation between energy saving due to the proposed actions and the related embodied energy;
- to evaluate the environmental benefits and drawbacks concerning the assessed retrofit actions.

Cumulative Energy Demand Method is used to account the primary energy requirement of the assessed actions (CED) (Frischknecht et al. 2007). The environmental impact assessment is carried out by means of indicators and characterization factors of the EPD scheme (Environmental Product Declaration–EPD) (EPD 2008).

13.3.2 Definition of the Functional Unit and System Boundaries

According to the UNI EN ISO 14040 standard, the functional unit is defined as the reference unit through which the performance of a product system is quantified in a LCA. In LCA studies of buildings the functional unit can be defined by its service time or by a reference area. In the examined case study each retrofit action proposed is selected as functional unit (FU) as following:

- FU_1: Thermal insulation of the building façade (224 m^2 width) by means of EPS board (Expanded Polystyrene) coating, 12 cm thickness. The thermal transmittance decreases from 0.96 to 0.27 W/m^2K;
- FU_2: Thermal insulation of the roof (142 m^2 width) by means of rock wool boards, 8 cm thickness. The thermal transmittance decreases from 0.60 to with a thermal transmittance of 0.25 W/m^2K;
- FU_3: Dismantling and renovation of the ground floor (142 m^2 width), adding a layer of XPS (Extruded Polystyrene), 8 cm thickness. The thermal transmittance decreases from 1.60 to with a thermal transmittance of 0.39 W/m^2K.

No retrofit action for windows is assumed because currently they are characterized by suitable thermal performance (transmittance lower than 3 W/m^2K). The following life cycle stages are assessed: manufacturing process of the retrofit components, installation, and transports. The environmental impacts caused by the infrastructures are neglected. Therefore, the impacts of the construction of roads, trucks used to carry the construction materials are not taken into account. With regard to the use phase, the annual energy consumption of the building after the retrofit accomplishment is estimated. With regard to the dismantling phase, due to the lack of data, it is not taken into account. This involved that in the following assessment the contribution of the retrofit end-of-life on the renovated building eco-profile is not estimated.

13.3.3 Data Quality in Life Cycle Inventory (LCI)

Data on material production involved for the three retrofit scenarios were provided by the producer companies in the sector. Fuel consumption and air emissions from transportation were calculated, depending on the transport mode and the distance between sites of production and the site of building construction. Eco-profiles of electricity, fuels, materials and installed components derived from energy and material inputs were referred to (Frischknecht et al. 2007). The annual energy requirement for the building operation after the retrofit actions was calculated through TRNSYS (TRNSYS v. 16.1 2003).

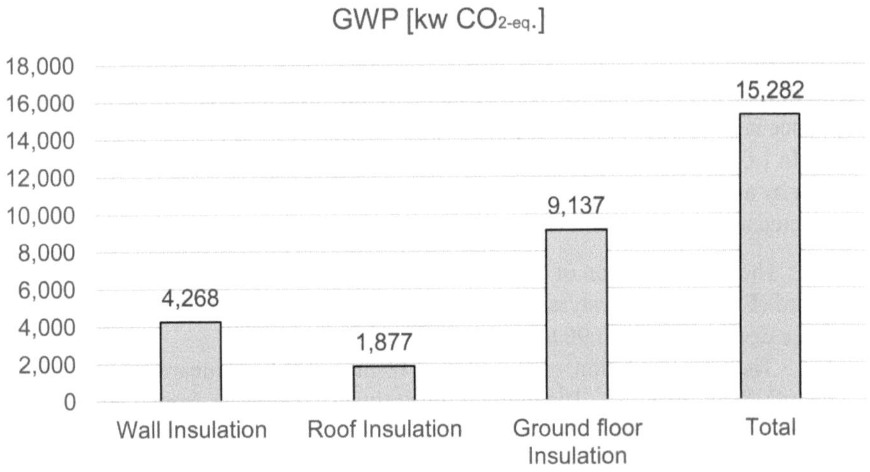

Fig. 13.2 Contribution of each retrofit options to Embodied energy, operating energy and demolition energy in the existing building

13.3.4 Life Cycle Impact Assessment (LCIA)

The highest contribution to the CED after the implementation of the retrofit actions comes from the production process of the involved materials for all the considered actions. The contribution of transport and installation can be neglected with regard to the production phase.

The highest contribution to the embodied energy is given by the retrofit of the ground floor (50 %), while the other actions shares for 32 % (wall insulation with EPS boards) and 16 % (roof insulation with rock wool boards).

Then the whole retrofit would increase the building embodied energy of 23 %.

The following environmental impact indicators are assessed, according to the EPD scheme: Global Warming Potential (GWP), Ozone Depletion Potential (ODP), Acidification Potential (AP), Nutrification Potential (NP), Photochemical Ozone Creation Potential (POCP).

Table 13.2 shows the results of the environmental impact assessment, according the EPD scheme (EPD 2008), while Fig. 13.2 shows the contribution of each retrofit option to GWP. The outcomes of the LCIA highlight that for each impact category the most significant contribution comes from the renovation of the ground floor, due to the relevance of the action (dismantling and reconstruction). In particular it accounts for about the 60 % on GWP, while the retrofit actions on the shell walls and on the roof account for 28 and 12 %, respectively. With regard to POCP, the renovation of the ground floor contributes for the 42 %, while the thermal insulation of the shell wall and the roof account for 25 and 33 % respectively.

With regard to AP and NP, the renovation of the ground floor contributes for the 42 and 40 % respectively; the wall retrofit involves higher contribution (34 and 33 %) than the roof retrofit (24 and 27 %). ODP is mostly due to the ground floor retrofit

Table 13.2 Environmental impact indicators per functional unit

	GWP100 (kg CO_2eq)	ODP (kg CFC_{11}eq)	AP (kg SO_2eq)	NP (kg PO_4^{3-}eq)	POCP (kg C_2H_4eq y)
EPS boards (FU1)	4,268	0.0003	16.62	1.57	0.62
Rock wool boards (FU2)	1,877	0.0005	11.61	1.31	0.80
XPS layer (FU3)	9,137	0.0510	20.83	1.98	1.04

Table 13.3 Embodied and operating energy before and after the assessed retrofit measures on the building envelope

	Existing building (GJ)	Renovated building (GJ)	Life cycle increase/saving (GJ)	Annualised increase/saving (GJ/y)
Embodied energy	1,217	1,436	+219	+4.4
Operating energy	3,327	2,865	−462	−9.2

13.4 Energy and Environmental Benefits Related to the Assessed Retrofit Scenarios

If all the retrofit measures were implemented the whole retrofit would imply an extra embodied energy of 219 GJ (20 GJ/m^2), thus increasing the building embodied energy to 1,436 GJ (13 GJ/m^2). The operating energy would decrease to 2,865 GJ (Table 13.3). The consequent variation in the building CED would be a net reduction of 243 GJ.

In order to assess the energy and environmental effects of the retrofit actions, the previous life cycle impacts are compared with the saved primary energy and the avoided environmental impacts, related to the reduction of electricity and fuel consumptions for cooling and heating (Table 13.4).

As showed in Table 13.4, the whole retrofit would involve energy and environmental benefits. Only ODP increases, essentially due to the resource consumption in the production processes of the insulation boards. The highest contributions to energy saving and avoided GWP arise from the insulation of the shell wall and of the ground floor, each one accounting for about 40 %.

The retrofit on the roof brings a small net energy saving, comparison with the other actions. Because of the small assumed improvement of the thermal performance for such a shell component, the related life-cycle impacts are nearly equal to the related energy and environmental benefits. Energy and GWP payback times are calculated for the whole retrofit.

The energy payback time can be defined as the time necessary for saving the primary energy equivalent to the CED of the retrofit. GWP payback time represents the time necessary to avoid the GWP equivalent to that one involved during the life cycle of the retrofit. If all the three retrofit actions were implemented, both the energy payback time and GWP payback times would result about 20 years.

Table 13.4 Annualized net energy saving and environmental benefits related to each retrofit action

	Annual avoided impacts					
	Energy	GWP (kg CO_2 eq/y)	ODP (kg CFC_{11} eq/y)	AP (kg SO_2 eq/y)	NP (kg PO_4^{3-} eq/y)	POCP (kg C_2H_4 eq/y)
Wall	1.93	143.40	−4.5E-06	0.0005	0.0220	0.461
Roof	0.90	76.84	−9.6 E-06	0.0004	0.0006	0.221
Ground floor	2.03	109.60	−1.02E-03	0.0005	0.0287	0.584
Total	4.86	329.84	−1.03 E-03	0.0014	0.0513	1.266

13.5 Conclusions

The following main hot-spots can be highlighted from the above LCA study:

- The LCA study of the existing building pointed out that the operation step involves the highest contribution to the life cycle primary energy consumption, accounting for 72 % of the CED. The monitoring of the user behavior during the operation step showed that the building annual operating energy mostly arises from the electricity consumption for lighting, electrical appliances and summer cooling, followed by the energy consumption for heating and DHW.
- The outcomes of the LCA of the retrofit actions, aimed at improving the thermal performance of the existing building, showed that these actions would involve additional primary energy consumption and environmental impacts.
- Looking at the building eco-profile as whole, the CED results decreased: the operating energy was reduced for heating and cooling, but it has still the highest contribution; an extra embodied energy of about 15 % is involved for the production and installation of the retrofit actions. Due to the lack of data the effect of the retrofit end-of-life on the building life-cycle was not accounted.
- Looking at the energy and GWP payback times, the obtained results revealed that the energy and environmental benefits would repay the retrofit life cycle impacts in a long period in comparison with the expected life-time of the retrofit actions themselves. Such a long period depends essentially on the relevant CED during the production of the synthetic insulation boards assumed foreseen in the retrofit actions.

References

Ardente F, Cellura M, Fontana M, Longo S (2008) Energy and environmental analysis of a mono familiar Mediterranean house. World Sustainable Building Conference SB08, Melbourne

Ardente F, Beccali M, Cellura M, Mistretta M (2011) Energy and environmental benefits in public buildings as a result of retrofit actions. Renew Energy Sustain Reviews 15:460–470

Blengini GA, Di Carlo T (2010) The changing role of life cycle phases, subsystems and materials in the LCA of low-energy buildings. Energy Build 42:869–880

Directive (2009) 2009/125/EC of the European Parliament and of the Council (2009) establishing a framework for the setting of ecodesign requirements for energy-related products (recast)

Directive (2010) 2010/31/EU of the European Parliament and of the Council (2010) on the energy performance of buildings (recast)

Dixit MK, Fernández-Solís JL, Lavy S, Culp CH (2010) Identification of parameters for embodied energy measurement: a literature review. Energy Build 42:1238–1247

Frischknecht R, Jungbluth N, Althaus HJ, Doka G, Dones R, Heck T, Hellweg S, Hischier R, Nemecek T, Rebitzer G, Spielmann M (2007) Overview and Methodology Ecoinvent Report No. 1, ver. 2.0, Swiss Centre for Life Cycle Inventories, Dübendorf (CH)

Gustavsson L, Joelsson A (2010) Life cycle primary energy analysis of residential buildings. Energy Buildings 42:210–220

Hernandez P, Kenny P (2011) Development of a methodology for life cycle building energy ratings. Energy Policy 39:3779–3788

Italian Decree n. 311 (2006) Disposizioni correttive ed integrative al decreto legislativo 19 agosto 2005 n. 192, recante attuazione della direttiva 2002/91/CE, relativa al rendimento energetico nell'edilizia, Gazzetta Ufficiale Italiana 01/02/2007 (in Italian language)

Sartori I, Hestnes AG (2007) Energy use in the life cycle of conventional and low-energy buildings: a review article. Energy Build 39:249–257

Suzuki M, Oka T (1998) Estimation of life cycle energy consumption and CO_2 emission of office buildings in Japan. Energy Build 28:33–41

TRNSYS v. 16.1 (2003) A transient system simulation program. Solar Energy Laboratory, University of Wisconsin

The International EPD Cooperation (2008), General Programme instructions for Environmental product declarations EPD. Version 1.0

UNI EN ISO 14040 (2006) Environmental management–Life cycle assessment–Principles and framework, International Organisation for Standardisation

Chapter 14
Bio-construction and Renewable Raw Materials: The Case of Cork

Alessio Tola

Abstract In the field of bio-construction the use of low environmental impact materials is increasingly widespread. Such components act as a factor of "passive climate conditioning" and require little energy during the production, maintenance and disposal/recycling phases. A great attention was granted to the principles of "social, mental and physical well being supported by the communities in favor of non-harmful materials", that discourage the use of non-transpiring insulating compounds of synthetic origin. Fiberglass or rockwool, polystyrene, urea foams, formaldehyde and other materials normally used in building can release potentially hazardous substances. Furthermore, their structure renders the insulated rooms waterproof affecting the transpiration; this, together with the sealing of doors and windows, contributes to indoor pollution. Bio-construction prefers to use thermal and acoustic insulators that are easy to use and to apply to supporting and/or covering structures, and to ornamental components as well. In the form of insulating panels of various thicknesses and densities, but also in agglomerates with other materials such as rubber, and the preparation of premixes, cork has a number of important applications in the construction sector. The various segments of the construction market (bio-construction, insulation and reconstruction of traditional buildings, large-scale works) constitute an important outlet for cork end-products, capable of promoting the diversification of the sector with regard to cork stoppers, expanding opportunities for operators in the sector and allowing the exploitation of main process waste products. The chapter analyses the production of sustainable insulating materials starting from the cork industry waste, with specific focus on the cork production district of Sardinia.

Keywords Bio-construction · Renewable · Cork · Insulate · Innovative

A. Tola (✉)
Department of Economic Sciences, University of Sassari, 07100 Sassari, Italy
e-mail: tola@uniss.it

R. Salomone, G. Saija (eds.), *Pathways to Environmental Sustainability,*
DOI 10.1007/978-3-319-03826-1_14,

14.1 Introduction

Technology plays an essential role in analytically understanding the role of a business, from both the strategic and technical and operational viewpoints: indeed, it is an integral part of the strategic policy by which companies define both the creation of competitive advantage, and its long-term sustainment (Porter 1985).

Indeed, decisions associated with growth resulting from internal development, but more generally, the choices relating to production diversification, also have several defining and leading resources in the technological variables.

The role of technology is essential in the most modern businesses, since it closely defines all corporate products and processes: the products and services in a company's portfolio are based on specific technological skills, so much so that each product can usually be broken down to the fundamental technologies possessed by a company.

Hence, technological innovation may also be defined as a novel combination of productive factors allowing the attainment of new products capable of supporting the beneficial expectations associated with exploitation of the same resources.

Innovation can be fed by scientific production cycles implemented through technological applications, or by economic-financial demands which, by supporting specific industrial scenarios and investing in certain applications, promote the affirmation of certain innovations.

In any case, independently of the developmental direction, it may be maintained that business opportunities are strongly linked to technological opportunities, which depend on one's knowledge at a given moment in time.

At the corporate level, technology represents a predominant share of the overall corporate infrastructure: indeed, it shapes all value chain primary and support activities, being central to all strategic choices.

The corporate product portfolio development process is inextricably linked to sustainable competitive advantage based on the ability to generate awareness and innovation, namely, to dynamically alter the technology of reference (Grando et al. 2006).

Hence, for businesses, the development and launch of new products and services have become activities that are absolutely essential for the purposes of their continued growth and improvement (Grando et al. 2006).

Within the manufacturing diversification process implemented by companies, the restraints resulting from the peculiarities of the raw materials used, the technological structure of the sector of reference and the socio-political framework of the market in which the companies operate define the majority of strategic choices.

Within the industrial manufacturing framework associated with the use of raw materials such as wood and cork, primary processing waste constitutes one of the emerging critical aspects: indeed, the defectiveness, or more simply, the characteristics of the materials themselves imply large quantities of process wastes.

The need to use process wastes in high added value secondary processes has always been an aim of production companies of this type, with a tendency to generate integrated manufacturing processes within the company or at the manufacturing sector level, so as to optimize the use of raw materials, in turn creating a diversified product portfolio capable of enhancing their own production, even in completely different markets.

This chapter underlines how the re-use of the waste material produced during the cork products manufacturing process could be a determining factor of sustainable energy efficiency development.

14.2 The Sardinia Cork Production District

The northern Sardinia province of Olbia-Tempio, and the towns of Calangianus and Tempio Pausania in particular, are associated with the production and processing of cork, with a history dating back to the ancient past. Cork-exploiting manufacturing activities have led to the creation of a true industrial district.

The District was only formally and legally recognized in 1997, following implementation of the legal prescriptions of Italian Law 317/1991. The district has a high level of entrepreneurial activity, significant levels of specialized manufacturing, but also high levels of interconnection between the companies present, demonstrating the development of a local system where integrated production cycles allow businesses to optimize processing wastes, even recovering the dusts produced by major manufacturing activities to generate energy (Tola 2012).

Territorially, the Cork Industrial District includes seven Alta Gallura towns, with a resident population of approx. 30,000 inhabitants, with a total of 100 businesses, 20 of which industrial in nature, processing over 20,000 metric tons of cork (including raw material and semi-finished product) on average, generating an overall turnover in excess of € 200 million.

Over the last 30 years, this concentration of businesses has produced direct employment for around 1,400 workers, in addition to approx. 1,200 employed in secondary activities. Unfortunately, the drop in employment levels over the past two years, first of all due to a crisis in the sector followed by the global crisis resulting from the financial crisis, has yet to be measured.

The main production output of the sector, closely bound to the availability of raw materials, so much so as to import over 50 % of the cork processed, is mainly aimed at the production of corks and stoppers[1] accounting for approx. 80 % of the turnover.

The manufacturing cycle generates a range of cork-processing waste products in excess of 70 % of the total raw material processed and has, as always, given rise to the age old problem of the beneficial exploitation of these residues (see Fig. 14.1) (Tola 2012).

In particular, with regard to cork discarded at the initial selection stage due to it being unsuitable for the cork manufacturing process (due to defects in the source plants, any disease in the plants that might result in the production of defective corks, and other reasons), this is not used in the production of corks for use in the

[1] The term corks is used to mean those stoppers obtained from a single piece of natural cork, while stoppers may also consist of those obtained by joining two or more pieces, from agglomerates of various granulates and by joining granulates with cork rings.

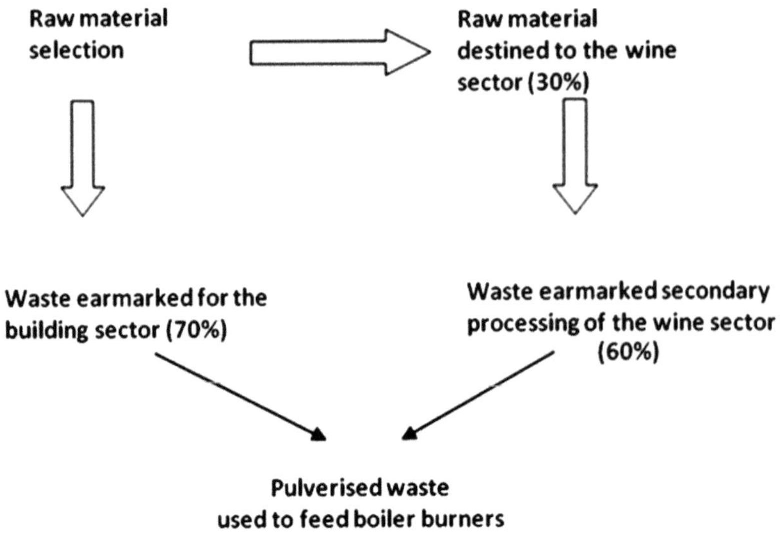

Fig. 14.1 Cork: raw material flow-chart. (Source: in-house data)

wine industry, but is transformed into granulate to be used in other manufacturing sectors. On the other hand, cork production wastes from the production of corks may be transformed into granulate for the production of agglomerate stoppers (Tola 2012).

The cork products obtained by granulate agglomeration may be classified into two major groups:

- Agglomerates for use in the wine industry;
- Agglomerates for use in the construction sector.

Indeed, research and innovation have permitted the use of the agglomerates in the construction sector, with different applications such as thermal and acoustic insulation and even decorative wall coatings.

14.3 Cork Products for Use in Construction and Other Applications

Building activity is one of the sectors with the highest environmental impact in terms of continuous and incessant use of the land, high energy consumption and the emissions associated with it. Building activities represent the largest consumer of energy European Union, accounting for 40 % of total consumption, and is the largest producer of CO_2 by representing 33 % of the total.

The issues on proper use of energy resources are therefore a matter of great importance and of great relevance, as confirmed by the targets set by the Kyoto

Protocol. The construction sector has long been at the heart of measures aimed at meeting EU commitments to reduce by 20% by 2020, global emissions of greenhouse gases and consumption of primary energy (Torretta 2011).

The environmental impacts associated with the "life cycle" of buildings are in fact investigated, in compliance with the regulations on voluntary and mandatory energy certification[2], aimed to obtain the Ecolabel and/or certification of other standards.

Specifically in Italy, ENEA shows that the buildings would need a specific energy consumption per square meter by far lower than that of other developed countries, but on the contrary, they mark specific consumption levels per square meter and degree-day among the highest. This suggests that, in Italy, the low power per square meter are due to the mild climate (geographical averaging of degree-days lower than 2,000) but our homes are poorly insulated and equipped with heating plants that are not properly managed.

As a biodegradable material, cork is preferable to the majority of artificial materials in a variety of applications ranging from stoppers, packaging and insulation, in addition to applications as a flooring material, since allergies associated with carpets are increasingly deemed a risk to health.

The role played by cork in the field of thermal insulation should be considered in the light of the extraordinary physical and technological properties said material possesses, in addition to its resistance to ageing and the action of rodents, insects and ice etc.

The structure of cork itself, consisting of dead cells filled with gas, very regular in size and numbering approx. 12 million cells per cubic centimeter, explains the heat transfer resistance properties. The cell walls are formed from three layers of tissue: a cellulose layer, a lignin layer and a suberinic layer.

While the first two layers predominantly perform a support role and are common in other plants, on the other hand, the suberinic layer is formed by a mixture of ethers made from long-chain fatty acids and alcohols. Hence, the suberinic layer is rich in fats and waxes, substances with waterproofing action, so much so that cork can remain submerged in water for years without losing its ability to float: indeed, the cell wall is impermeable to water, and few solvents are capable of passing through (Teece 1997).

The major applications in the construction sector relate to the use of this material in the form of insulation panels of various thicknesses and densities, but also as a material available in various granulometries, used in combination with traditional or innovative construction products (slaked lime and liquid rubbers) (see Table 14.1).

The market for cork in the construction industry may be broken down into various sectors, including:

- Bio-construction;
- Traditional buildings (insulation and renovations);
- Large-scale building works.

[2] Directive 2002/91/CE of the EU Parliament and of the Council, December 16, 2002, on energy efficiency in buildings (in OJ L January 4, 2003, no. 1).

Table 14.1 Insulating materials: main characteristics. (Source: Bureau of energetic efficiency, Autonomous Province of Bolzano (Italy))

	Thermal conductivity λ in W/mK	Equivalent thickness [1] (cm)	Resistance to diffusion coefficient μ	Raw material availability	Energy required for production	Environmental pollution during production	Energy required for transport	Recycling
Calcium silicate	0.05–0.07	12–17	6	Abundant	High	Not listed	Low	Rarely possible
Expanded perlite	0.01–0.06	10–15	1–4	Abundant	Medium	Low	Medium	Reassemblable
Expanded polystyrene (EPS)	0.035–0.04	9–10	20–100	Limited	High	High	High	Rarely possible
Extruded polystyrene (XPS)	0.035–0.04	9–10	80–200	Limited	Very high	Very high	High	Rarely possible
Linen	0.04	10	1	Reproducible	Low	Low	Medium	Reassemblable
Rockwool or fiberglass	0.035–0.04	9–10	1–2	Abundant	Medium	Medium	Low	Reassemblable
Hemp	0.04	10	1	Reproducible	Low	Low	Low	Reassemblable
Wood fiber	0.04	10	5	Reproducible	High	Medium	Low	Reassemblable
Cork	0.04	10	15–18	Reproducible	High	Low	Medium	Rarely possible
Expanded mineral	0.045	11	5	Abundant	High	Medium	Low	Rarely possible
Polyurethane (PUR)	0.025–0.03	6–8	30–100	Limited	High	Very high	High	Rarely possible
Lambswool	0.04–0.045	10–11	1–2	Reproducible	Low	Low	Low-high	Reassemblable
Cellular Glass	0.04–0.05	10–12	watertight	Abundant	High	Medium	Medium	Rarely possible
Cellulose	0.04	10 cm	1,5	Recycled product	Low	Low	Medium	Rarely possible

[1] Thickness of insulating material equivalent to 10 cm of insulating material with λ=0.04/mK

The first two sectors envisage the use of cork as a thermal and/or acoustic insulating material, while in the third it is used for elastic-expansion and vibration-proof joints.

Regarding the classification of cork products for the construction industry, the following macro-classes may be identified:

- Cork granules;
- Cork granules premixed with specific construction components;
- Rigid and flexible cork agglomerates for insulation and decorative applications.

The first type deals with products resulting directly from the grinding of processing wastes, suitably sorted in relation to granulometry, to be used in cavities without any further processing.

With regard to premixes, these are essentially distinguished as:

- *Blocks*: consisting of cork granules, natural hydraulic lime, crack-resistant fibers and natural additives making the product: heat-insulating, sound-absorbent, transpiring, de-humidifying, de-icing. The main uses envisage applications in new buildings for insulating and sound-proofing both flat and sloping surfaces, floorings, and blocks in general, and in the renovation of old buildings for the purposes of restoring and dehumidifying blocks and flooring.
- *Plasters*: With the same basic components as the blocks, but with finer granulometry cork, allowing the protection of interior and exterior walls from rising damp, condensation and infiltration (Rubini 2009).
- *Special products* made from cork micro-granules with additives for sealing holes and packing.

The following classification is proposed for the various types of cork agglomerate for insulation and decorative applications:

- *Rigid or flexible raw granulate cork agglomerate.* The raw granulate based rigid product is very well suited to use as a vibration-proof insulation material and for housing tiles. This type of agglomerate is produced starting from raw granulate, *i.e.* obtained from seasoned, unboiled cork, originating from planks of cork unsuitable for transformation into corks, and from virgin cork. The granulate is supplemented with other components in order to improve resistance to the action of hot and cold water and potential fungal or bacterial attack; in addition, they preserve the natural color of the cork. The flexible product, characterized by the presence of particular resins capable of endowing the product with elasticity and flexibility, may be used as insulating tape for preventing the dispersion of heat from pipes containing hot fluids.
- *Rigid or flexible boiled granulate cork agglomerate.* With regard to the types of product resulting from the use of boiled granulate, it should be said that these are generally made using smaller granulometry softened white granulates. The methods for preparing the blocks are the same as for the rigid type of product obtained from raw granulate.

- *Cork flour agglomerate*. This product is flexible and slightly spongy. It may be used as insulating tape and, underneath standard tiles, as thermal and acoustic insulation, even at thicknesses of just a few millimeters, both for the attenuation of sound transmitted through the air and for avoiding condensation. At the same time, it can be a decorative element, replacing standard wallpapers, with the advantage of being able to be painted and serviced in general.
- *Agglomerated cork foam*. This is obtained by cooking raw cork granulate in autoclaves using a superheated steam system (approx. 300 °C): it is the best product for thermal insulation; As a result of the high pressures and temperatures, the cork expands and expels the waxes and resins which then solidify on contact with air, cementing the granules together. As soon as the high pressures are removed, the blocks obtained in the autoclave undergo a rapid recovery, making the surfaces of the blocks themselves unsuitable for direct use. Trimming of the blocks gives rise to additional waste which will be reused for the production of black regranulates.
- *Black regranulate agglomerates*. These are obtained from the regranulation of agglomerated cork foam processing wastes. Adhesives are added to the granulate in order to obtain insulating panels that are more flexible than rigid agglomerated panels.
- *Glued cork foam*. This is obtained by gluing black regranulate using synthetic adhesives. Rigid and flexible products, with varying degrees of compactness, may be obtained. The compressed, rigid product may be used for the applications envisaged in sections 1 and 2 and as a coating for pipes with diameters in excess of 10 cm. This type of coating is very cheap since it can be produced in appropriate shapes, *i.e.* without processing wastes.

14.4 Sustainable Economic Development in the Construction Sector

The concept of sustainable development represents an extremely important turning point in the traditional way of perceiving the relationship between economics and improving environmental quality: it is a new development model aimed at transforming the traditional economics-environment negative correlation into an integrated and positive relationship, by relying on novel production methods and new strategies envisaging economic development through the rational exploitation of natural resources while safeguarding quality of life.

This concept has been brought to the attention of public opinion and scholars for the first time in the report of the World Commission on Environment and Development in 1987 (*Brundtland Report*, United Nations, 1987), which has been applied in recent years, especially with the 1997 Kyoto Protocol coming into effect on 16 February 2005 (Italy ratified the protocol with Law No. 120 of 1 June 2002), envisaging the obligation for Industrialized Countries to reduce greenhouse gas emissions by 5.2 % on average, with respect to 1990 levels, within the period

2008–2012. Specifically, Europe must implement a reduction of 8 % with respect to 1990 levels of emissions, and Italy, which represents the third highest emitting country in the EU after Denmark and Spain, must implement a reduction of 6.5 %.

However, only in recent years and somewhat late, Italy registered a reduction in emissions, a further demonstration of the uncoupling between economic growth and energy consumption.

Certainly, sustainable construction, the main objective of which is the improved energy and environmental performance of the building-installation system, can contribute towards achieving the above-indicated aims through:

- The use of renewable energy sources;
- Saving primary resources;
- Defining novel construction technologies;
- The use of eco-compatible materials;
- The introduction of confined space air-conditioning systems that take into account the thermal insulation and the efficiency of the installations.

Italy has witnessed rapid scientific and technological development thanks to:

- The increased sensitivity of the population to environmental matters;
- The search for and identification of innovative technological solutions for containing the high energy running costs for existing buildings;
- The implementation of initiatives aimed at defining guidelines for the evaluation/certification of buildings and providing operators in the sector with a better understanding of the long-term costs and the direct and indirect benefits of sustainable construction;
- Reinforcing the manufacturing fabric in terms of eco-innovation.

Regulatory constraints, closely associated with reducing energy consumption resulting from application of the Kyoto protocol, allow promotion of the principles of saving energy and consequently the use of eco-efficient technologies within the framework of the thermal insulation of buildings.

The aim of strategies targeted at building energy savings, promoted throughout Europe in recent years by research and regulatory bodies, is reducing the dispersion of heat from buildings. For the purposes of reducing energy consumption, design attention is frequently focused on reducing the loss of heat through the shell and, consequently, on increasing the level of thermal insulation; in a well insulated building, the heat delivered to a room by the heating system is trapped and not released to the outside.

14.5 Conclusions

Increased sensitivity to environmental matters can lead to promotion of the use of cork materials in construction applications, having considered the good functional characteristics of said products.

Attention to environmental aspects in product and material usage cycles, assessment of the overall cost of the usage cycles of products themselves, but especially the altered definition of the analysis of the costs and the value of property over the long term brings sustainability and usage variables of low environmental impact materials to the forefront.

The market development strategies of such types of products must obligatorily pass through such actions as:

support for the concepts of environmental sustainability and reduced use of non-renewable resources in the construction sector;

integration between the companies producing environmentally sustainable materials and potential customers;

training of personnel to improve organizational efficiency and the ability manage new market opportunities and strategies aimed at sustainability;

Public Administration policies aimed at greater communication of technologies for the adoption of innovative materials aimed at lowering the energy costs of manufacturing facilities and buildings.

References

Grando A, Verona G, Vicari S (2006) Tecnologia Innovazione Operations [Technology innovation operations]. Egea, Milan

Porter ME (1985) Competitive strategy. Creating and sustaining superior performance. Free Press, New York

Rubini R (2009) Gli utilizzi del sughero in applicazioni innovative [Uses for cork in innovative applications]. In: Meeting report—Innovazione e valorizzazione del sughero [Cork Innovation and Development], Tempio Pausania, Italy

Teece DJ, Pisano G, Shuen A (1997) Dynamic capabilities and strategic management. Strat Manag J 18(7):509–533

Tola A (2012) Innovazione tecnologica, ecosostenibilità e sviluppo competitivo nel settore del sughero [Technological innovation, ecosostenibility and competitive development in cork sector]. Franco Angeli, Milan

Torretta P (2011) Costruire sostenibile: necessario e conveniente. http://www.icmq.it/download1.php?f=DOWNPUB_28itfqaz. Accessed 26 Nov 2012

Chapter 15
Studies About the Utilization of Citrus Wastes in View of Environment Protection

Francesco Lanuzza, Fabio Mondello and Maria Marcella Tripodo

Abstract The increasing needs to protect environment require a correct management of wastes, including agro industry wastes deriving from citrus processing. These industries produce in the world an amount of wastes estimated in about 14 million t every year. This chapter concerns the utilisation of citrus industry wastes as peels and refining pulps of the juices. From these wastes it is possible to obtain products employed in the alimentary and pharmaceutical industries, as pectin, mucilage, flavonoids and feed for animals. In particular, lemon peels are mostly utilized in order to obtain good quality pectin, instead orange peels and juice pulp are used for production of fodders. From lemon juice pulp it is possible to draw mucilage and flavonoids. Furthermore citrus wastes may be treated and valued in order to obtain derivatives utilizable as source of energy, that are bioethanol and biogas obtained by fermentation and other materials useful by physical treatments. Therefore, the procedures described appear attractive in economic terms adding value to this waste and beneficial to the environment reducing the microbiological contamination too.

Keywords Citrus waste utilization · Citrus pectin · Lemon mucilage · Hesperidin recovery · Citrus waste feed · Energy from citrus waste

15.1 Introduction

The increasing needs to protect environment require a correct management of wastes, including citrus industry wastes. Today, these wastes are increasingly regarded as good secondary raw materials for obtaining high value substances or as feedstuff. Furthermore, if these wastes are not used or not adequately disposed off, they can contribute to environmental pollution. It is common knowledge that citrus fruit are eaten as fresh fruit or they are processing to producing juices and essential

F. Lanuzza (✉) · F. Mondello
Department of Economics, Business, Environment and Quantitative Methods,
University of Messina, Piazza S. Pugliatti 1, 98122 Messina, Italy
e-mail: francesco.lanuzza@unime.it

M. M. Tripodo
Department of Chemistry Sciences, University of Messina, 98166 Messina, Italy

R. Salomone, G. Saija (eds.), *Pathways to Environmental Sustainability*,
DOI 10.1007/978-3-319-03826-1_15,

Table 15.1 Citrus world production—year 2010 (tons)

	Orange	Lemon and lime	Mandarin tang. clem.	Grapefruit (inc. pumelos)	Total
Brazil	18,101,700	1,020,350	1,122,730	72,100	20,316,880
China	5,003,289	1,058,105	10,142,430	2,884,820	19,088,644
USA	7,477,920	800,137	540,682	1,123,100	9,941,839
India	5,966,400	2,629,200	–	260,600	8,856,200
Mexico	4,051,630	1,891,400	409,442	–	6,352,472
Spain	3,120,000	578,200	1,708,200	43,200	5,449,600
Italy[a]	2,486,131	567,597	841,589	7,575	3,902,892
Turkey	1,710,500	787,063	858,699	213,768	3,570,030
Egypt	2,401,020	318,111	796,867	–	3,515,998
Argentina	833,486	1,113,380	423,737	188,820	2,559,423
Iran (Islamic Rep. of)	1,502,820	706,800	276,138	46,500	2,532,258
South Africa	1,414,590	215,985	142,500	343,055	2,116,130
Pakistan	1,505,000	–	559,000	–	2,064,000
Indonesia	2,028,900	–	–	–	2,028,900
Morocco	849,197	–	472,834	–	1,322,031
Other	3,464,400	1,237,676	2,415,259	1,519,091	8,636,426
Total	61,916,983	12,924,004	20,710,107	6,702,629	102,253,723

Data processing by the authors on FAOSTAT 2010

[a] Data from ISTAT 2010

oils. The wastes of citrus industry are polluting because are wet (over 80%), rich of sugars and of other compounds that make it fermentable.

Worldwide citrus fruit production in 2010 amounted to 102,253,723 t (see Table 15.1), including 60% of oranges (FAOSTAT 2010). On 2012 the Italian production reaches 3,682,000 t of which 45% obtained in Sicily (ISTAT 2012). On average Europe converts about 20% of oranges compared with 70% of Brazil and 75% of USA (USDA 2012). The Italian citrus fruit industry processes approximately 40% of the oranges and 65% of lemons (SG Marketing Agroalimentare 2011).

From data USDA (2012) world orange juice production during marketing year 2011/2012 in selected major-producing countries is estimated at 2.1 million t (65 °Bx), so the worldwide industrial citrus peel wastes may be estimated in about 14×10^6 t. Citrus peel wastes are partly used for cattle feed either fresh or after dehydration, but considerable quantity of citrus pulp in the fresh state is lost as a result of the difficulty of rapidly disposing a large quantity of the wastes in a fairly short period of time, then a huge amount of it is still discarded to nature, causing several environmental problems. So, the disposal of these citrus wastes poses a serious problem for citrus industries around the world (Talebnia et al. 2008).

In Europe the development outlook of citrus industry is limited due to competition from South American countries, USA, and nowadays from Asian countries such as China and India. European industry therefore needs to find rational and economically viable ways to exploit the intrinsic potential in citrus fruit. This means making use of the solid residue besides producing juice and essential oil. The solid residue is mainly peel but also contains pulp, seeds and membranes, on average accounting for over 50% of the fruit in weight (Braddock 1999).

In the current industrial practice, citrus processing waste (CPW) is first pressed to remove free liquid, the expressed liquor evaporated, added back to the press cake, and then dried to be sold as cattle feed to avoid waste disposal cost. The dried CPW (10% moisture) is composed of sugars (30–40%), pectin (15–25%), cellulose (8–10%) and hemicellulose (5–7%) (Grohmann et al. 1995).

In Italy this process is not much utilized considering the high cost of fuel and also for the few advantage of this processing in little manufacturing firms.

The authors of this chapter have been engaged in research to fully exploit the various waste products of citrus fruit (De Gregorio et al. 2002; Lanuzza et al. 2002; Lo Curto et al. 1992; Micali et al. 2002; Tripodo et al. 2002; Tripodo et al. 2004). Other studies treat of the exploitation of the citrus wastes to get fibres and food ingredients as pectins and mucilages (Ben-Shalom and Pinto 1999; Davidson and McDonald 1998). However, as regards the lemon, in the zones of its production and manufacturing, among which Spain and Italy, the peel wastes are actually used as feed for livestock but mainly for the industrial production of pectin.

Besides the peel, another waste product that may be generated by concentrated juice producers is the "refining pulp" of the juice. This is the sediment generated from centrifuging the juice (mostly of orange). It is a semi-solid material with water content in excess of 85% and cannot therefore be dried as such in normal industrial driers. As this material is rich in sugars, fibre, flavonoids and other organic substances, it is not easy to treat in normal waste treatment plants. The quantity of refining pulp, named also puree for the typical consistency, can total about 10 t every day in a middle manufacturing firm.

Pectin and mucilage are substances with high value added that may be obtained from citrus wastes. Pectin are natural products of vegetable origin, interesting as dietary supplements and also for the cosmetic and pharmaceutical industries due to their gelling, thickening and stabilizing properties (Mesbahi et al. 2005; Piriyaprasarth and Sriamornsak 2011). They are polysaccharides and can be classified as "dietary or food fibre". Mucilages are soluble dietary fibres that are present in some citrus waste and other vegetables. They are also vegetal polysaccharides similar to pectin but have a different composition in uronic acids and sugars (Karawya et al. 1980). Like pectin, they can be used by food, dietary, cosmetic and pharmaceutical industries; recent studies have highlighted the anti-inflammatory effects of lemon mucilage (Galati et al. 2005).

Other interesting products that can be obtained from citrus waste are flavonoids, recognized as antioxidants and widely found in fruit and vegetables (El Nawawi 1995). The same are utilized both in the food and pharmaceutical fields. In particular, hesperidin can be extracted from lemon and orange peel and is used in the pharmaceutical industry for its anti-inflammatory and vasodilatory properties (Da Silva et al. 1994; Galati et al. 2005).

Furthermore citrus wastes may be treated and valued in order to obtain derivatives utilizable as source of energy by fermentation, and other useful materials by physical treatments.

Another alternative to improve the management of these residues is the implementation of new processes for their recovery, for instance, through the production of organic fertilizers or as a substrate for the production of several compounds with high value added, such as microbial proteins, organic acids, enzymes and biologically active secondary metabolites and adsorbent materials (Rezzadori et al. 2012).

15.2 Citrus Wastes as a Source of Products with High Value Added

15.2.1 Orange Peel

Lo Curto, Tripodo et al. (1992) proposed a process on orange peel utilization. From the solid residue obtained after an acidic pre-treatment of the peel, hesperidin, in larger quantities and at lower cost than for a traditional method, may be recovered. A continuous fermentation for SCP (Single Cell Protein) production, utilizing *Geotrichum candidum*, was carried out on the pre-treatment liquid.

As the hesperidin content in the peel is of the order of 2–3 % on dry matter, to obtain 1 kg of this product 200–250 kg of wet peel must be processed. Consequently large quantities of valueless solid wastes as well liquid effluents with polluting characteristics are obtained, whose disposal could be a problem.

In this process the flavonoid is extracted from the solid residue obtained after pre-treatment. Therefore is used less raw material and the flavonoid yield is doubled. The yield of hesperidin from the pre-treated orange peel ranged between 3.7 and 4.5 % on dry matter, whereas it ranged between 1.8 and 2.3 % when extracted from fresh orange peel. The better yield is due to the solubilisation of pectin after the acidic pre-treatment as above said. Moreover, the waste materials present minor problems because the final liquid obtained after flavonoid separations contains few fermentable substances and the new solid residue may be utilized (after drying) as a feedstuff. In fact it shows an interesting composition: 4–5 % crude protein, 24–28 % crude fibre and 73–78 % in vitro digestibility. It should be suitable as a feedstuff for ruminants and also no-ruminants, because of its quite low content of fibre.

As far as the SCP characteristics are concerned, the analysis revealed that the crude protein content ranged between 39 and 44 %; crude fibre between 4.8 and 6.1 %; ash between 7.1 and 7.5 %, total in vitro digestibility rose to 75–85 %.

15.2.2 Orange Pulp

The authors of this chapter describe a method which makes it possible to obtain animal feed from orange juice centrifugation pulp (Tripodo et al. 2004). To this end, alkaline (CaO) and/or enzymatic (Citrozym LS) treatments were carried out on the centrifugation pulp. These treatments facilitate pressing and so help to produce a material which, using suitable methods, may be dried.

Enzyme treatment proved to be the most efficient of the methods under investigation designed to favour the pressing of the pulp.

The product obtained with this method showed excellent digestibility in vitro (88.6%) and its protein content (7.7%), although not especially high, compared favourably with that of many other agro industrial waste products currently used as components of animal feed.

As a by-product of the pressing of the pulp, a liquid is obtained which, after concentration in a multi-step concentrating device, can be recycled by sending it back through the pressing sequence again alongside a new batch of treated pulp. Thanks to the multi-step concentration technique (for example, using three sequential evaporators), about one third less heat is required to evaporate off the water from the filtrate obtained during pressing. When the concentrated liquid is reintroduced into the production cycle, it enriches the treated pulp with soluble substances and so further enhances the nutritional value of the final product. The steam produced during the concentration process could be used as a recyclable energy source within the production process itself.

Finally, in addition to the clear environmental benefits to be gained from the recycling of the pulp from centrifuged orange juice, the citrus processing industry might also receive a technological and economic boost from the introduction of these new production cycles to work alongside their traditional ones.

15.2.3 Lemon Peel

Lemon peels are employed, as said above, for industrial production of pectin. The authors of this chapter proposed a new process for the simultaneous production of pectin and hesperidin, a flavonoid with interesting properties (Tripodo et al. 2012). This simultaneous extraction method utilizes chemical reagents, as calcium hydroxide, hydrochloric acid and ethanol. With this process it is possible to get two fraction of pectin: pure pectin separately from a low grade pectin with a lesser degree of polymerization. By comparing the yields with separated extractions of pectin (21.61 %) and hesperidin (0.51 %), the simultaneous extraction process gave higher rates for both pectin (22.80 %) and hesperidin (0.76 %). Obviously, the figure for pectin was compared with the total amount of pectin obtained by summing the "pure" and "low grade" pectin with the single yield obtained from the process to extract pectin alone. Although the yield for hesperidin is not high in absolute terms, in view of its high value added and the fact that it is obtained as a secondary product from pectin production, its recovery is equally interesting in economic terms. Furthermore, the characterization analyses [content of protein (7.8 %), crude fibre (19.8 %), ashes (11.6 %), digestibility (60.3 %)] of the final material, obtained after extraction process, show that they could be suitable for use in animal feed.

The proposed procedure not only enable substances of high value added to be obtained but also allow the final waste to be recovered and used in animal feed.

15.2.4 Lemon Pulp

Similarly about the process of lemon peels, the authors of this chapter proposed for pulps also a simultaneous drawing of mucilage and hesperidin (Tripodo et al. 2012). The employed reagents are calcium hydroxide, hydrochloric acid and ethanol, as above, but in different concentration and in different conditions (temperature and contact times). Once again, it is evident that the simultaneous extraction procedure affords higher yields (mucilage 17.50 %, hesperidin 1.48 %) than extracting the single products (mucilage 16.48 %, hesperidin 1.01 %). The mucilage yields are of interest given the high value added of this product. Hesperidin yields are about one and half time those obtained from the peel. This is not surprising given that the source of the pulp is the centrifuged juice, which is richer in hesperidin. Also in this matter, characterization analyses [content of protein (8.9 %), crude fibre (18.7 %), ashes (14.9 %), digestibility (84.8 %)] allow to state that the final residue could be an utilizable component for feedstuff, just like other food industry wastes. The same could be employed for ruminant animals but also for no ruminants (e.g. chicken and pigs) because of its good digestibility and correspondent low fibre content.

An alcoholic solution is obtained as a by-product of mucilage production, and this is distilled to recover the ethanol. The waste water solution, after concentration in a multi-step concentrating device, could be added to the final residue and thereby reintroduced into the production cycle. In this way, the concentrated liquid enriches the final residue with soluble substances, and so further enhances its nutritional value.

15.3 Citrus Wastes as a Source of Energy

The citrus wastes also may be treated and valued in order to obtain derivatives utilizable as source of energy. These wastes can have high energy value, and they can help to reduce the dependency on energy purchased for the generation of heat, vapor, or electricity.

15.3.1 Bioethanol and Biogas

Citrus wastes can be used to produce ethanol, limonene, and other co-products (Martin et al. 2010). Limonene is employed in the manufacture of food and medicines as a flavouring agent and has many applications in the chemical industry as well as in cosmetics and domestic household products.

Extensive work has been done with enzymatic hydrolysis to liquefy and maximize monomeric sugar content. These sugars can be subsequently or simultaneously converted into ethanol by fermentation (Boluda-Aguilar et al. 2010; Grohmann et al. 1994; Grohmann et al. 1998; Wilkins et al. 2007a; Wilkins et al. 2007b). Widmer et al. (2010) studied the use of citrus processing waste for ethanol production by saccharification and fermentation processes and observed that the ethanol yields based on the sugar content after enzymatic hydrolysis following 48 h of simultaneous saccharification and fermentation ranged from 76 to 94%. As limonene is an inhibitor of yeasts, removal of limonene is a critical step in the production of ethanol from citrus wastes.

An alternative solution can be anaerobic digestion (in which both pollution control and energy recovery can be achieved), mainly of orange peel waste. This process is defined as the biological conversion of organic material to a variety of end products including biogas whose main constituents are methane (65–70%) and carbon dioxide (Rezzadori et al. 2010). Martín et al. (2010) have shown that the anaerobic digestion of orange peel waste, after a d-limonene extraction, reveals higher methane production, providing an excellent opportunity to integrate this waste into orange juice manufacturing using a biorefinery approach.

An integrated process can be used to produce ethanol, biogas and limonene from orange wastes (Pourbafrani et al. 2010). Depending on the market and the profitability of the process, pectin can also be recovered as a by-product. Pourbafrani et al. suggested the application of this new process to produce ethanol and biogas and provided a configuration for its industrial application. Their proposal is based on the simplicity of the process and low price of the biomass compared to other ethanol processes involving the use of lignocelluloses, making this process unique and favorable. Citrus wastes are hydrolyzed by dilute-acid process in a pilot plant reactor equipped with an explosive drainage. By applying this process, the same authors refer that 1 t of citrus wastes with 20% dry weight resulted in 39.64 L ethanol, 45 m^3 methane, 8.9 L limonene, and 38.8 kg pectin. Lohrasbi et al. (2010) carried out the economic analysis of this process. Recovery of limonene improves the economy of the process and makes citrus waste-to-ethanol process more competitive than a cellulose-to-ethanol process. In this process, the steam requirement is fulfilled by burning 29% of the produced methane, which results in low utility cost compared to the total operating cost.

15.3.2 Bio-oil and Charcoal

Despite the great opportunities for the use of such citrus wastes, there are few examples in the literature of the application for energy generating purposes, either in the form in which they are produced or after their transformation (Boluda-Aguilar et al. 2010; Widmer et al. 2010). According to Haykiri-Acma et al. (2010) and Rutkowski (2011), the biomass used as an energy source is mainly a mixture of three main components, i.e., cellulose, hemicellulose and lignin.

Due to the high content of these components in the orange waste, the residue generated from orange juice production is considered a good biomass for the production of bio-oil.

Although research in this area is still in the initial stages, bio-oil has great potential to replace petroleum, with advantages in various applications. The bio-oil is a renewable organic fuel derived from the processing of agricultural and forest residues. It is a black oil with a characteristic odor and high performance and can be obtained not only from pyrolysis but also thermal degradation. Bio-oils may be used for many applications for example as a transport fuel or as an additive to conventional fuels and to feedstock chemicals (Rezzadori et al. 2012).

Fast pyrolysis processes are the most commonly used for the production of bio-oil. This process converts thermo-chemically ground particles of biomass into bio-oil. It also produces bio-charcoal, which can be used as a fertilizer, even in organic production systems, and many other applications, such as the prevention of environmental pollution and decontamination of water bodies and soil affected by toxic metals. The use of waste orange (peel and pulp) for energy is of great interest since every 100 kg of such waste can provide around 27 kg of charcoal (Rezzadori et al. 2012).

15.4 Conclusions

Between citrus fruit, oranges are mostly produced and processed all over the world, after follow lemons. Into about 14 million t of citrus industry wastes obtained every year, over 12 million t come from oranges and 1 million from lemons, the remaining comes from other citrus, as mandarin, grapefruit, etc. Considering the big quantity, from orange processing waste it could be more advantageous to obtain feed, after drying, and/or energy product (mainly ethanol and biogas) through fermentation processes. Instead, from lemon processing waste it could be more favourable to produce substances with high value added, as pectin and mucilage. The reason of above is determined from the fact that lemon peels contain more pectin (over 20 % on dry matter) than orange peels (about 16 % on dry matter) and further lemon pectin is of better quality owing a longer chain. These processes reduce, not at all but meaningfully, waste quantity to dispose of.

Production costs of ethanol and biogas probably are not advantageous for selling, however if they are used in the same firm they may reduce energy costs.

About the Italian situation, which citrus industries are little-medium sized, it is obvious that there are unfavourable conditions to produce energy materials from citrus wastes. In view of the above, solution of this matter should be a sole bigger plant setting up in the area. In the end, the citrus waste can be a resource for the same firm or can be reused as a raw material for other industries, adding value to this residue and reducing the microbiological contamination and the environmental pollution as well as the cost of exploration of natural resources as mineral bio-oil and ethanol obtained from non-renewable sources.

References

Ben-Shalom N, Pinto R (1999) Natural colloidal particles: the mechanism of the specific interaction between hesperidin and pectin. Carbohydr Polym 38:179–182. doi:10.1016/S0144-8617(98)00111-8

Boluda-Aguilar M, García-Vidal L, del Pilar González-Castañeda F, López-Gómez A (2010) Mandarin peel wastes pretreatment with steam explosion for bioethanol production. Bioresour Technol 101:3506–3513. doi:10.1016/j.biortech.2009.12.063

Braddock RJ (1999) Handbook of citrus by-products and processing technology. Wiley, New York

Da Silva Emim JA, Oliveira AB, Lapa AJ (1994) Pharmacological evaluation of the anti-inflammatory activity of a citrus bioflavonoid, hesperidin, and the isoflavonoids, duartin and claussequinone, in rats and mice. J Pharm Pharmacol 46:118–122. doi:10.1111/j.2042-7158.1994.tb03753.x

Davidson MH, McDonald A (1998) Fiber: forms and functions. Nutr Res 18:617–624. doi:10.1016/S0144-8617(98)00111-8

De Gregorio A, Mandalari G, Arena N, Nucita F, Tripodo MM, Lo Curto RB (2002) SCP and crude pectinase production by slurry-state fermentation of lemon pulps. Bioresour Technol 83:89–94. doi:10.1016/S0960-8524(01)00209-7

El Nawawi SA (1995) Extraction of citrus glucosides. Carbohydr Polym 27:1–4. doi:10.1016/0144-8617(95)00036-7

FAOSTAT (2010) Statistical database of the Food and Agriculture Organization of the United Nations. http://faostat.fao.org/site/339/default.aspx. Accessed 08 Oct 2012

Galati EM, Cavallaro A, Ainis T, Tripodo MM, Bonaccorsi I, Contartese G, Taviano MF, Fimiani V (2005) Anti-inflammatory effect of lemon mucilage: in vivo and in vitro studies. Immunopharmacol Immunotoxicol 27:661–670. doi:10.1080/08923970500418919

Grohmann K, Baldwin EA, Buslig BS, Ingram L (1994) Fermentation of galacturonic acid and other sugars in orange peel hydrolysates by the ethanologenic strain of Escherichia coli. Biotechnol Lett 16:281–286. doi:10.1007/BF00134626

Grohmann K, Cameron RG, Buslig BS (1995) Fractionation and pretreatment of orange peel by dilute acid hydrolysis. Bioresour Technol 54:129–141. doi:10.1016/0960-8524(95)00121-2

Grohmann K, Manthey JA, Cameron RG, Buslig BS (1998) Fermentation of galacturonic acid and pectin-rich materials to ethanol by genetically modified strains of Erwinia. Biotechnol Lett 20:195–200. doi:10.1023/A:1005349112770

Haykiri-Acma H, Yaman S, Kucukbayrak S (2010) Comparison of the thermal reactivities of isolated lignin and holocellulose during pyrolysis. Fuel Process Technol 91:759–764. doi:10.1016/j.fuproc.2010.02.009

ISTAT—Italian National Institute of Statistics (2010) Agriculture and livestock. http://agri.istat.it/sag_is_pdwout/jsp/NewDownload.jsp?id=15A|21A|31A & anid=2010. Accessed 05 Dec 2010

ISTAT—Italian National Institute of Statistics (2012) Agriculture and livestock. http://agri.istat.it/sag_is_pdwout/jsp/NewDownload.jsp?id=15A|21A|31A & anid=2012. Accessed 05 Dec 2012

Karawya MS, Wassel GM, Baghdadi HH, Ammar NM (1980) Mucilages and pectins of Opuntia, Tamarindus and Cydonia. Planta Med 40:68–75. doi:10.1055/s-2008-1075006

Lanuzza F, Micali G, Coppolino R, Tripodo MM (2002) Recupero di flavonoidi dalla polpa di raffinazione dei succhi di arancia (flavonoids recovery from orange juice pulp). In: Proceedings of the Euroconference on University and Enterprise—a partnership for training, research, employment and social development, University "La Sapienza", Rome, 26–28 September 2002

Lo Curto RB, Tripodo MM, Leuzzi U, Giuffrè D, Vaccarino C (1992) Flavonoids recovery and SCP production from orange peel. Bioresour Technol 42:83–87. doi:10.1016/0960-8524(92)90065-6

Lohrasbi M, Pourbafrani M, Niklasson C, Taherzadeh MJ (2010) Process design and economic analysis of a citrus waste biorefinery with biofuels and limonene as products. Bioresour Technol 101:7382–7388. doi:10.1016/j.biortech.2010.04.078

Martín MA, Siles JA, Chica AF, Martín A (2010) Biomethanization of orange peel waste. Bioresour Technol 101:8993–8999. doi:10.1016/j.biortech.2010.06.133

Mesbahi G, Jamalian J, Farahnaky A (2005) A comparative study on functional properties of beet and citrus pectins in food systems. Food Hydrocoll 19:731–738. doi:10.1016/j.foodhyd.2004.08.002

Micali G, Lanuzza F, Tripodo MM, Coppolino R, Mondello F (2002) Valutazione della polpa di raffinazione degli agrumi come sorgente di fibre dietetiche (valuation of citrus pulp as a source of dietary fiber). Proceedings of the Euroconference on University and Enterprise—a partnership for training, research, employment and social development, University La Sapienza, Rome, 26–28 September 2002, pp 783–790

Piriyaprasarth S, Sriamornsak P (2011) Flocculating and suspending properties of commercial citrus pectin and pectin extracted from pomelo (*Citrus maxima*) peel. Carbohyd Polym 83:561–568. doi:10.1016/j.carbpol.2010.08.018

Pourbafrani M, Forgács G, Sárvári Horváth I, Niklasson C, Taherzadeh MJ (2010) Production of biofuels, limonene and pectin from citrus wastes. Bioresour Technol 101:4246–4250. doi:10.1016/j.biortech.2010.01.077

Rezzadori K, Benedetti S, Amante ER (2012) Proposals for the residues recovery: orange waste as raw material for new products. Food Bioprod Process 90:606–614. doi:10.1016/j.fbp.2012.06.002

Rutkowski R (2011) Pyrolysis of cellulose, xylan and lignin with the K_2CO_3 and $ZnCl_2$ addition for bio-oil production. Fuel Process Technol 92:517–522. doi:10.1016/j.fuproc.2010.11.006

SG Marketing Agroalimentare (2011) Gli Agrumi. Produzione e superficie 2004–2010. http://www.sgmarketing.it/public/download/111_AGRUMI_Produzioneesuperficie_2010.pdf. Accessed 07 July 2011

Talebnia F, Pourbafrani M, Lundin M, Taherzadeh MJ (2008) Optimization study of citrus wastes saccharification by dilute-acid hydrolysis. Bioresources 3:108–122

Tripodo MM, Lanuzza F, Micali G, Coppolino R, Nucita F (2004) Citrus waste recovery: a new environmentally friendly procedure to obtain animal feed. Bioresour Technol 91:111–115. doi:10.1016/S0960-8524(03)00183-4

Tripodo MM, Lanuzza F, Micali G, Nucita F (2002) Valorizzazione di scarti agricoli e agroindustriali mediante estrazione di sostanze a medio-alto valore aggiunto: mucillagini e pectine (exploitation of agricultural and agroindustry wastes by extraction of high value products: mucilages and pectin). In: Morselli L (ed) Proceedings of "Seminari RICICLA", Maggioli Ed. SpA, Rimini, 6–9 November 2002

Tripodo MM, Mondello F, Lanuzza F (2012) Simultaneous recovery of dietary fibres and hesperidin from industrial lemon waste. Forum Ware 2:23–30

USDA—United States Department of Agriculture, Foreign Agricultural Service (2012) Citrus: world markets and trade. http://www.fas.usda.gov/psdonline/circulars/citrus.pdf. Accessed 04 Oct 2012

Widmer W, Zhou W, Grohmann K (2010) Pretreatment effects on orange processing waste for making ethanol by simultaneous saccharification and fermentation. Bioresour Technol 101:5242–5249. doi:10.1016/j.biortech.2009.12.038

Wilkins MR, Suryawati L, Maness NO, Chrz D (2007a) Ethanol production by Saccharomyces cerevisiae and Kluyveromyces marxianus in the presence of orange-peel oil. World J Microbiol Biotechnol 23:1161–1168. doi:10.1007/s11274-007-9346-2

Wilkins MR, Widmer WW, Grohmann K (2007b) Simultaneous saccharification and fermentation of citrus peel waste by *Saccharomyces cerevisiae* to produce ethanol. Process Biochem 42:1614–1619. doi:10.1016/j.procbio.2007.09.006

Chapter 16
Prospects for the Use of Mini and Micro Wind Power Plants in Residential Settings

Salvatore Chiricosta, Simona Saccà and Giuseppe Saija

Abstract The use of mini and micro wind turbines in residential settings is one element of the ever widening use of small-scale, sustainable and eco-compatible electricity generation systems. This type of machine has been greatly improved in recent times, taking on very interesting technical characteristics which mean this sector is expected to experience rapid growth over the next few years. This chapter gives a brief overview of the use of wind power for the production of electricity and discusses the main technical and economic aspects linked to the use of small scale wind turbines in residential settings, with particular reference to the situation in Italy.

Keywords Sustainable development · Microwind turbines · Miniwind turbines · Residential setting · Energy saving

16.1 Sustainable Development and Wind Power

The sustainability of the current global economic-production system, which has been severely weakened by the recent economic crisis, is under serious threat from growing energy requirements, particularly from developing countries. This trend is having serious repercussions both on an environmental level, with the growing risk of real climate change with potentially destructive and irreversible effects, and on an economic level, with continual riskiness of supplies and energy price variability.

In order to make economic development sustainable there is a need to make drastic changes, in as short a time as possible, in the direction of efficient, accessibly priced, low-carbon energy supplies (Frankl 2010).

Renewable energy sources, particularly wind power, obviously play an essential role in the mix of low-carbon energy technologies.

S. Chiricosta (✉) · S. Saccà · G. Saija
Department SEAM, University of Messina, Piazza S. Pugliatti 1, 98122, Messina, Italy
e-mail: salvatore.chiricosta@unime.it

R. Salomone, G. Saija (eds.), *Pathways to Environmental Sustainability,*
DOI 10.1007/978-3-319-03826-1_16,

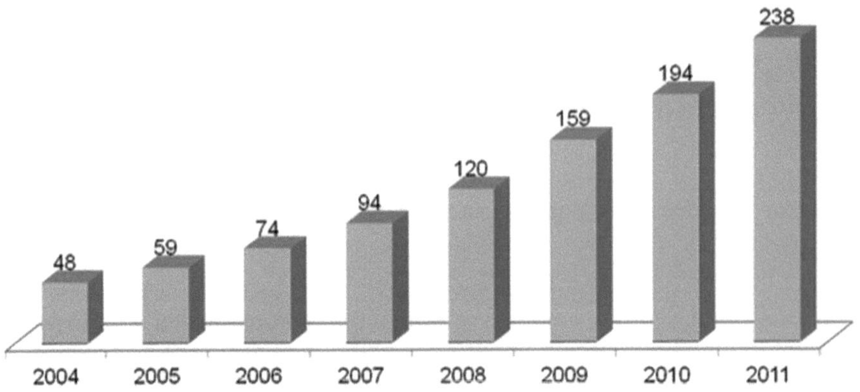

Fig. 16.1 Accumulated global wind power (expressed in GW) from 2004 to 2011. (Source: Own elaboration of GWEC data)

16.1.1 Wind Power Globally

Wind power, despite the current crisis, recorded a new record in 2011, with 43,961 MW of new power installed, globally, and an increase of about 22.6 % compared to 2010. Total power has thus reached an accumulated world total of 238 GW (Fig. 16.1). Traditionally, the countries that have invested and produced the most in the wind power sector have been Germany, Spain and the USA. Since 2005, however, China has arrived on the international scene, making increased investments in this sector. Thanks to the 18,000 MW installed in 2011 (40 % of the world total), China has succeeded in consolidating its lead in just 6 years: 62,733 MW, 26.3 % of total global wind power. China is followed by the USA, with 6,810 MW installed in 2011, making an overall total of 46,919 MW, leaving behind Germany (+2,086 MW in 2011, 29,060 MW overall) and Spain (+1.050 MW in 2011, 21,674 MW overall). After these four nations (which alone account for 67.3 % of global wind power capacity), India is gaining ground (+3,019 MW in 2011, 16,084 MW overall); on the other hand, France (+830 MW in 2011, 6,800 MW overall) and Italy (+950 MW in 2011, 6,747 MW overall) are lagging behind, closely followed by Great Britain (+1,293 MW in 2011, 6,540 MW overall) and Canada (+1,267 MW in 2011, 5,265 MW overall), which, on the contrary, are experiencing rapid growth (GWEC 2011). This means that 86.5 % of global wind power is installed in just ten countries: China, the USA, Germany, Spain, India, France, Italy, Great Britain, Canada and Portugal (EWEA 2012).

It is estimated that, by 2020, total wind power capacity could reach: 587,000 MW, providing about 6 % of the world's electrical power; or, according to more optimistic forecasts, with adequate political support, it could reach 1,100,000 MW, covering 11.7–12.6 % of global electricity requirements (GWEC and Greenpeace 2012).

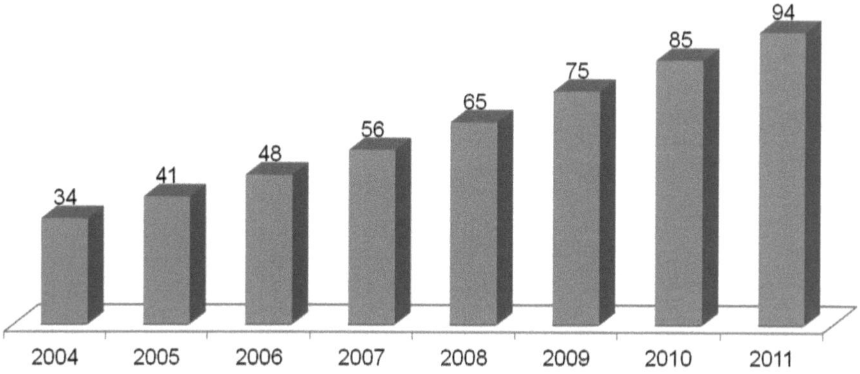

Fig. 16.2 Accumulated European wind power (expressed in GW) from 2004 to 2011. (Source: Own elaboration of EWEA data)

16.1.2 Wind Power in Europe

The 27 member states of the European Union (EU) are responsible for 42 % of global wind power production. In 2011, about 9,600 MW of wind power capacity were installed, reaching a total of about 94 GW (Fig. 16.2), sufficient for the production of 204 TWh of electricity and for covering 6.3 % of electricity consumption in the EU. The European wind power industry experienced an average annual growth rate of 15.6 % over the period 1995–2011 and now represents 21.4 % of renewable energy production. According to the latest statistics published by EWEA (European Wind Energy Association), the EU has installed more than 100 GW of wind power, enough to satisfy the requirements of 57 million houses. This means that the wind can be used to produce a quantity of electricity equal to that produced by 39 nuclear power plants or 62 coal fired or 52 gas fired plants. This corresponds to an annual CO_2 emission reduction of about 219.5 Mt in the event of wind power replacing coal, and of about 97.8 Mt of CO_2 in the event of replacing gas (EWEA 2012).

16.1.3 Wind Power in Italy

From the late 1990s onwards, the spread of wind power has led to Italy, with 6,936 MW installed at the end of 2011, becoming the fourth European country in terms of power installed, behind Germany, Spain and France. Figure 16.3 shows the accumulated power and the number of plants installed in Italy from 2004 to 2011 (GSE 2012).

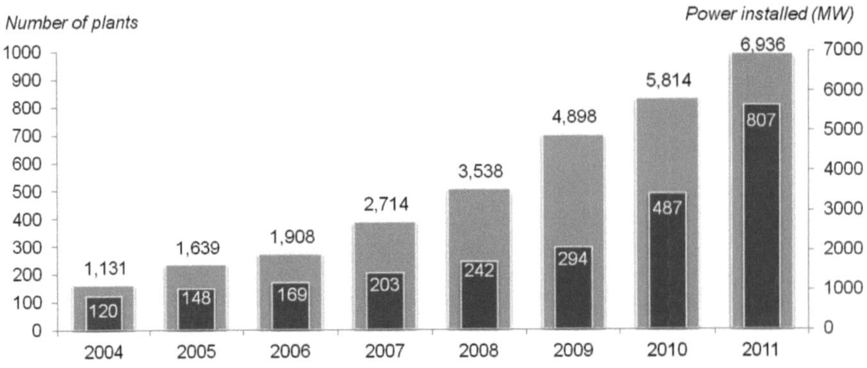

Fig. 16.3 Accumulated wind power (expressed in MW) and plants installed in Italy 2004–2011 (Source: Own elaboration of GSE data)

With regard to the total amount of electrical power produced from wind sources, Italy is 13th[1] in Europe, having produced 9,126 GWh in 2010, covering 2.85% of overall demand. This sector is in continuous expansion, to the extent that there were as many as 807 plants installed in 2011. Due to particular environmental and territorial characteristics (windiness, orography, accessibility), most of these are situated in the southern regions (98% of national wind power and 80% of the total number of plants). The region producing most energy from wind power is Sicily, with 2,370 GWh, followed by Puglia (2,256 GWh) and Campania (1,344 GWh). These three regions account for 61% of the national total (GSE 2012). 450 Italian municipalities have wind power production plants; of these, 346 have achieved energy independence thanks to wind power (Legambiente 2012).

16.2 Small Scale Wind Power Plants

Large wind turbines have now reached a level of development at which their technology is often defined as "mature". However, the possibility of fully exploiting wind power must not only be pursued with the building of "wind farms" but also through the creation of small scale wind power plants. These were initially neglected because of their insufficiently wide production scale, but are now acquiring growing importance in the residential sector, since there are considerable advantages to their use, such as negligible acoustic and electromagnetic impact, practically no interaction with resident and migratory bird life, and moreover, their limited dimensions do not bring any unpleasant changes to the esthetic perspective of the landscape. To this we need to add the highly accessible costs, low maintenance expenses and, even more importantly, the possibility of decentralizing electricity production and making it available to everybody (Paraggio 2005). Thus, while pho-

[1] Denmark is the leading country in Europe, covering 25.9% of its national energy requirements with wind power.

tovoltaic energy has so far been predominant in residential settings, a certain degree of interest is now being registered in aerogenerators for domestic use which transform wind energy into power, with values varying from less than a hundred watts to several dozen kilowatts.

Currently, there is no universally recognized classification of small scale turbines; one possibility is that of using the size limits laid down by the International Electrotechnical Commission (IEC) 61400 Standards. In particular, IEC Standard 61400-2[2] (Wind turbines—Part 2: Design requirements for small wind turbines), of which the second edition has been published and which has also been implemented in Italy as CEI EN 61400-2, classifies "mini turbines" those which are for residential, rural and workshop use (Community WT) which have a rotor swept area $\leq 200\ m^2$, corresponding to a nominal maximum power level of 100 kW; "micro turbines" are defined as domestic turbines (Home WT) with a rotor swept area $\leq 2\ m^2$, corresponding to a nominal maximum power level of 0.5 kW (Battisti 2012).

Nonetheless, there are "virtual distinctions" between micro and mini turbines, tied to specific power thresholds. In fact, considering Italian legislation, the 2008 budget law (Law 24/12/2007 no. 244) lays down that only plants of <200 kW can take advantage of the incentives for the all-inclusive tariff and, thus, plants with a power level below this threshold are classified as "mini wind turbines". On the other hand, "micro wind turbines" are specified as being those with very low power levels, generally ≤ 20 kW. Plants below this threshold do not require registration as an "electrical workshop" with the U.T.F. (*Ufficio Tecnico di Finanza*—Technical Finance Office); therefore, a plant of such small dimensions does not have the right to sell the energy produced, which is only for self-consumption, and, as regards authorizations, it is subject only to the *Segnalazione Certificata di Inizio Attività*—Certified Notice of Activation (SCIA) (Consulente-Energia.com 2011).

16.2.1 Aerogenerators for Mini and Micro Wind Power Plants

As for turbines of larger size, both mini and micro turbines use the classical "horizontal axis" and "vertical axis" aerogenerators. The former, which are simply smaller versions of larger ones, are best suited to open spaces, while the latter, with vertical rotors, are more suitable for urban environments and for urban wind patterns, which are usually turbulent, with winds of varying strength and direction.

"Vertical axis generators" are of completely different shape and size to classical rotors, with the aim of exploiting the force of the wind with very low levels of visual and acoustic impact. Moreover, they do not need to be oriented, since the surface area exposed to the wind is a 360° arc, and may be of interest for use in buildings with roofs not necessarily south-facing and on small terraces. This is the category with the most interesting ideas for small scale users and, judging from the variety of products on offer, it is expected to be a sector with a strong rate of growth, stimu-

[2] The third edition of the standard is currently in CDV (Committee Draft for Vote) form and is expected to be issued early in 2013.

lated by incentives and tax relief policies adopted in certain countries, with the aim of developing an alternative energy model to the traditional one. The positive assessment of this kind of plant is demonstrated by the fact that worldwide mini wind power installation has increased, in terms of power, from 105.9 MW, in 2006, to 275.8 MW, in 2010. The three largest world markets for mini wind turbines are the USA, the UK and Canada, although reliable data is currently lacking with regard to China, India and Japan (Bardi 2012).

An American company based in California, Prevailing Energy Inc., specialized in the high technology generation of wind power, has developed and is marketing the first electric generator based on a vertical multiple rotor blade wind turbine, called *Zephyr VATG* (Vertical Axis Turbine Generator). This turbine is made of composite material in fiberglass and aluminum, so as to reduce its weight (ensuring greater wind sensitivity) without reducing the resistance of the structure: in practice the material used is the same as that used for helicopter rotor blades and can resist wind strengths up to 240 km/h. The rotor blades are of an elongated rectangular shape, and the design is based on the same aerodynamic concepts used in designing aircraft wings. Depending on the size of the turbine, Zephyr can develop a power level between 3 and 150 kW. The shaft of the Permanent Magnet Generator (PMG) is directly attached to the turbine (direct driveshaft) and this makes the system extremely reliable. The PMG produces a three-phase alternating current. The "stand alone" electronic part, called "ZONEWORK", is composed of un storage system from 6 to 48 V, a 220 V AC inverter and a charge controlled device which redirects power when the batteries are charged (Prevailing Energy 2012).

Another US company, Cleanfield Energy, has developed a new ultra silent model of mini turbine, capable of fully exploiting urban wind, even when it is scarce or inconstant. This new technology, called *Vertical Axis Wind Turbines* (VAWTs), could be tested in the city of Toronto in Canada thanks to collaboration, also of a financial nature, with the Ontario Centers of Excellence and technical support from the McMaster University. The system would avoid the need to build traditionally sized wind turbine towers in rural zones a long way from urban areas or of great environmental importance and to equip them with expensive systems for transferring the energy produced to the urban electricity grids. According to Cleanfield Energy, this is a potentially huge market and, once the project is approved, the turbines that will be installed could be V3.5 VAWT, the latest 3.5 kW model, which will also be exported to Europe, above all to Great Britain, Ireland and Slovenia, where Cleanfield already has business links (Tita Gallo 2010).

16.2.1.1 The Production of Aerogenerators in Italy

In Italy, despite the numerous problems still holding back the development of this segment, mini wind power has started growing in importance as is shown by the fact that over 13 MW and 300 plants were installed at the end of 2011 (9.1 MW in 2011 alone) and that a further 10 MW will be installed in 2012 (Buonfrate 2012). The Italian market offers a wide range of small sized aerogenerators, mostly of excellent quality. There are the "traditional" ones, equipped with the classical two or three rotor blades, and the more innovative ones, characterized by their more unusual shape designed to better exploit the force of the wind and to be less dangerous for birds. The average price varies a great deal, but the parameter that has most effect on the final cost is undoubtedly the power (expressed in watts) of the plant installed which is related to the width of the rotor (Casale et al. 2012).

The Ropatec T30pro is a vertical axis aerogenerator, derived from the hybrid H- Darrieus-Savonius turbine, but more sophisticated aerodynamically, produced by an Italian firm based in Bolzano, which is activated by the minimum of wind (2–3 m/s). The turbine is essentially made up of a rotor element and a stator element, joined by three low friction ball bearings. The generation of electricity is achieved through a three phase brushless permanent magnet generator (PMG). These are attached to the rotor component of the generator, while the generator windings are connected to the stator component. This construction design, lacking in mechanical parts, strongly characterizes the performance of the machine, in terms of acoustic noise levels, minimum wear and tear and electricity generation. The size of the machine is such that power levels cannot reach > 30 kWp; the installation of a brake allows the speed of the turbine to be limited to a maximum of 18–20 m/s (Ropatec 2012).

Another leading Italian green economy firm is *Enervolt*, based in Vicenza, which has a vertical axis turbine called 'Forza 7' as its leading product: this is a vertical mini wind power plant, available in 3 kW and 5 kW versions, made of stainless steel or coated aluminum resistant to any environmental conditions; moreover, in order to better comply with landscape restrictions, it is also available in transparent polycarbonate (Nonsolosolare 2009). The main advantages of this turbine are: extreme noiselessness allowing it to be installed either on roofs or also on appropriate steel staffs; the capability of producing energy whatever the wind direction; automatic start up at low speed; the possibility of increasing the power produced if arranged in parallel (Furlan 2012).

ADAG (Aircaft Design & AeroFlightDynamics Group) based at the University Federico II of Naples has designed and built Mythos, a turbine 5.8 m in diameter with a low starting speed (2.5 m/s), suitable for areas where traditional rotor blades would not be efficient. There is also a vertical axis version for installation on city buildings. This is a project from the Eolpower spinoff, protected by patent, produced in collaboration with other companies (Dello Iacovo 2011).

One of the principal firms, particularly for domestic mini wind turbines, is *Jonica Impianti* (a partner of *Energie Naturali Srl*), a leader in the wind aerogenerator sector. Its two best selling models are the *Jimp 30* and *Jimp 60*. The former aerogenerator, with a power level of 25/30 kW, can reach a height of 30 m and a weight of 900 kg and is composed of three rotor blades. Its main characteristics are: a rotor diameter of 10.8 m; a synchronous, multi-polar, axial flow, permanent magnet generator, with an output voltage up to about 380 V. The *Jimp 60*, on the other hand, with the same type f generator, achieves a nominal power level of 60 kW with a height of 36 m, a weight of 6,500 kg and a rotor diameter of 18 m (D'angelo 2011).

16.2.1.2 Recent Developments in Mini Wind Power

One of the new developments in mini wind power was presented at the World Climate Summit in Cancun (29 November–10 December 2010) by the Californian entrepreneur Dan Bates and his team. This is *TurboMill*, a vertical wind turbine of extremely limited dimensions, which can be installed on the roof of a house and connected by cable to a normal power socket. Thanks to this revolutionary invention, any power socket is potentially capable of being transformed into a small private generator, directly producing energy for the household. It performs well even in minimum wind conditions as it is equipped with a modular system of interconnected turbines so as to increase power levels. *TurboMill* is made up of three vertical axis turbines, about 1 m high and a little wider, with an overall power level of about 500 W and can generate energy even at very low wind speeds; the grid inverter is incorporated. *TurboMill* is: cheap; available in numerous colors; easy to install; silent; of low visual and environmental impact; suitable for urban or rural settings. It can be installed on a floor, roof or wall and is connected directly to a household power socket. The current price is around € 1,000 (Elesis Impianti ed Automazioni Srl 2012; Facciolla 2011).

A very recent project, still in its initial development phase but with great prospects following its favorable reception by the market, is that of creating transportable wind power; it is a micro wind generator called *Eolic Foldable*, which is foldable and transportable. The overall design allows the owner to have all the components needed for electricity production from wind power grouped together in one unit. The technical characteristics and energy production capacity are still unknown, but it is thought that this aerogenerator will be capable of satisfying the energy requirements of small temporary camps or guaranteeing the minimum needs of mountain huts or small chalets. The materials used will be aluminum and carbon, so as to make the structure particularly lightweight and flexible (De Nardo 2010).

16.3 Micro and Mini Wind Turbines in Italy

The brief discussion above cannot certainly claim to be comprehensive, given that there are over 150 firms operating in this sector on the Italian market, employing an average of 10–15 people, engaged in design and installation as well as production of aerogenerators and their components; however, it does highlight the growing interest shown in the micro and mini wind power market. Despite is high potential, estimated at 1 GW of installations in the long term, this market is still modest in size, with a parceled industrial supply chain and lacking in large scale businesses. The challenges facing Italian makers of small scale turbines in the next few years, in order to avoid this sector being dominated by foreign producers (mainly Chinese), are those of improving the technology and performance of their machines, which need to have very high levels of reliability because of their target users (Buonfrate 2012). Indeed, the turbines must have very limited maintenance requirements and adequate safety levels for operation in environments with considerable human presence, thus involving greater risk of damage to persons and objects in the event of breakage. This means carefully checking environmental compatibility, both in terms of acoustic noise levels and operational safety, as well as the capability of extracting power efficiently at the low wind speeds typically found in such complex environments (Battisti 2012).

In order to provide guarantees to this end, various initiatives have been launched, also on an international level. In December 2009 the AWEA (American Wind Energy Association) drew up a *technical standard* which can be used voluntarily to test small wind power systems from safety and performance viewpoints. Another organization, the *Small Wind Certification Council* concerns itself with certifying systems tested on the basis of this *technical standard.* The mini wind power industry has always considered certification as a desirable objective, seeing it as a means of providing consumers, regulatory bodies and political decision makers with precise, credible and transparent information on the safety, performance and reliability of this technology (ENEA 2009).

In Italy the company *ICIM S.p.A.*, based in Sesto San Giovanni (MI), has developed a certification scheme specifically for wind turbines, based on the CEI EN IEC 61400-1 and CEI EN IEC 61400-2 standards (valid for all power levels, from micro wind turbines to the largest and most powerful ones) and updated in late 2010 following the publication of the new IEC 61400-22 standard, which adds performance tests to those already required by existing standards (ICIM 2011). This company took the lead on certification of renewable energy with the first certificate of conformity to ISO EN 61400-1 in the wind power sector issued for the aerogenerator *Libellula 55 kW*, the flagship of an entirely Italian firm, *Aria srl* based in Prato (FI) (Aria srl 2012).

In order to evaluate the possibility of using small scale wind power plants, certain technical and economic aspects need to be considered.

To determine the potentially exploitable wind energy in a given area we need to know the installation site, wind direction and speed trends over time and its distribution at that height. Verifying windiness, especially in environments with a strong human presence (presence of buildings) and for aerogenerators with short supporting towers (as is the case for mini wind turbines) is extremely difficult and the information contained in wind atlases or provided by anemometric stations cannot always be effectively used (Microeolico.com 2012). Beyond this, it is indispensable to have other information needed to carry out an overall evaluation of the plant (purchase and installation cost, guarantees, maintenance, bureaucratic procedures, etc.). It may often be necessary to use a consultancy firm, experts in the sector, offering "turnkey" installation of the aerogenerator, following the customer through the four basic phases indicated in Table 16.1 (SEI—Società Elettrica Italiana 2012).

In general, the ideal position for an aerogenerator is a piece of land with a low degree of roughness and with a gradient between 6 and 16°. The wind must exceed a speed of at least 5.0 m/s and must blow constantly for most of the year. Obviously, the dimensions of the aerogenerator vary as power levels increase, as can be seen from Table 16.2, which also shows the characteristics of the aerogenerators on the

Table 16.1 The four basic phases necessary for the installation of micro and mini wind turbines

Consultancy	Planning	Installation	Assistance
Anemometric measurements Feasibility studies Anemological studies Funding applications	Certified notice of activation (SCIA) Environmental impact evaluation (V.I.A.) Application for incentives Single authorization Application for connection to the electricity grid	Supervision of works Safety schemes Aerogenerator installation	Ordinary maintenance Extraordinary maintenance Emergency intervention Monitoring

basis of their nominal power level (up to 200 kW) and their typical uses (Casale et al. 2012).

The wind power plant has a significantly higher construction and installation cost (closely tied to the anemometric measurements at the site) if compared with that of other renewable energy forms (photovoltaic, mini-hydroelectric). It can be estimated that a mini wind turbine with a power level between 20 and 60 kW has an average cost of between 3,000 and 5,000 €/kW installed (since there are fixed elements regarding the unit cost per kW, this will be higher for smaller plants than for those above 20 kW). When costs are evaluated, it is also necessary to consider that a wind power plant has an average life span of about 20 years with low maintenance costs (Enel.si 2009). Overall, it has been calculated that the time needed for a return on investments, according to the windiness of the site, vary from a minimum of 7–8 years to a maximum of 12 years (Giubilo 2012).

The spread of wind power may be encouraged by state incentives for energy savings, which have certainly contributed to the development of the market for renewable energy sources.

In Italy, a Ministerial Decree issued on 18 December 2008 incentivized wind power plants up to 200 kW with a very advantageous all inclusive tariff of 0.30 €/kWh net produced for a period of 15 years. The possibility of obtaining this incentive ended on 31 December 2012, the last available day for activating a wind power plant and have 15 years at this tariff. Moreover, a new decree on renewable energy was issued, which allows for a reduction of the incentive from 0.30 €/kWh net fed into the electricity grid to 0.291 €/kWh, for plants with power levels between 1 and 20 kW, and to 0.268 €/kWh for those with power levels > 20 kW and up to 200 kW. The incentive will no longer be valid for 15 years but will be extended to 20 years (GURI 2012). With these values, the reduction of the incentive is compensated for, in a certain sense, by the extension of the incentive payment period. The new all inclusive tariff, which will be launched in 2013, does not substantially modify the previous one, indeed it is an improvement for plants below 20 kW. There is a substantial benefit for smaller products and for use in residential settings, offering interesting returns on investment with a life span of at least 20 years.

On expiry of the incentive period, the extra electricity produced can be resold to the Energy Services Provider (GSE) by means of "feed-in"; or, by using "net metering", it can be recovered later as required (a meter measures the amount of

Table 16.2 Characteristics of aerogenerators, classified on the basis of their nominal power and their typical uses

Electrical power produced at nominal speed	Rotor diameter	Height of staff	Typical uses
A few hundred watts	1–2 m	2–6 m	Boats, campers, small isolated users
From 1 to 6 kW	2–5 m	6–8 m	Dwellings, shops and SMEs, ground or roof installations also in urban areas, isolated users or connected to the electricity grid
>6 up to 60 kW	5–18 m	8–30 m	Farm holiday centers, camp sites, villages, shops, farms and SMEs, for ground installations and users connected to the electricity grid
Over 60 up to 200 kW	18–30 m	30–60 m	Farms and SMEs, for ground installations and users connected to the electricity grid

electricity fed into the grid and the amount consumed and at the end of the year the two amounts are balanced). The regulations establish that up to 100 kW the connection to the electricity grid must be Low Tension (LT); sometimes also for higher power levels, but only if the grid allows it. Normally, all plants up to 60 kW are connected in LT, while those above 100 kW are connected in medium voltage, so as to guarantee the stability of the grid. For connecting to the grid, therefore, all mini wind plants deal with the local electricity distribution company (White Wind 2010).

However, looking to 2013, the growth prospects for mini wind power is more uncertain because of the introduction, over the next 3 years, of a 60 MW/year limit to the power available to plants between 60 kW and 5 MW. Within this power range, and thus excluding micro and mini plants < 60 kW, the plants will be subject to the mechanism of registration, a procedure already used for photovoltaic power, which will introduce a further bureaucratic complication from the authorization viewpoint and will create greater uncertainty with regard to the time needed for returns on investment, and further difficulties in access to credit (Buonfrate 2012).

Finally, one of the most serious problems faced by current wind power plants is the difficulty of disposal for many rotor blades at the end of their life cycle. Indeed, the majority of the rotor blades are made from composite materials which contain a great deal of epoxy resins, obtained from petrochemicals, and fiberglass which, at the end of their useful life, are very difficult to recycle. They are usually disposed of in dumps, burnt as fuel for electricity generation or broken up and used as fillers in construction. The laboratories of two American universities, the University of Massachusetts Lowell and the Wichita State University in Kansas, are developing an innovative method for the creation of new turbines using sustainable materials. In this way, thermosetting epoxy resins would be obtained from vegetable oil, a "green", non-toxic, sustainable and readily available raw material, thus reducing energy consumption and production costs to a minimum. This would also make it possible to recover and recycle these materials when the rotor blades are decommissioned (Aguirre 2012).

16.4 Conclusions

The development of the mini wind power market in Italy is a recent phenomenon and is still not widespread, despite the huge potential demonstrated by the great success of large scale wind power. This market is mostly made up of small businesses, both Italian and foreign, which produce components, build and install mini wind power plants with power levels up to 200 kW. Italy has excellent prospects in this sector, with a supply chain consisting of specialized mechanical firms which are now reconverting to mini wind power production. Moreover, the use of wind power for the production of clean and renewable energy is growing in importance as one of the major options for energy diversification, allowing the exploitation of wind energy in order to achieve two advantages: saving money and protecting the environment. We need only think that, unlike what happens with the use of fossil fuels, mini and micro wind turbines, installed all over the world, reduce global CO_2 emissions by 40 million tons. Moreover, while there were limits imposed until recently due to the invasiveness of the plants, in terms of height and dimensions, the new mini wind power technology has brought about a significant reduction of these obstacles.

Considering the current situation of great political-economic instability on a global level, it must be said that being able to rely on the capacity to self-produce electricity means significantly reducing national energy costs. Moreover, the spread of the use of mini and micro wind turbines in residential settings may: serve as protection against the ever more frequent risk of electrical black-outs; create greater choice for consumers, in terms of energy supplies; allow the creation of interesting job opportunities and the development of new professions in urban and rural areas. It is, thus, important for Italy to take advantage of this significant development opportunity and to aim to become the leading country in the field of renewable energy, both in Europe and in the Mediterranean area.

References

Aguirre EL (2012) Innovative turbine blade material will be derived from vegetable oil. http://www.uml.edu/News/stories/2011-12/Niezrecki-NSF-grant.aspx. Accessed 2 Oct 2012

Aria srl (2012) http://www.aria-srl.it/it/aria-srl-en. Accessed 15 Feb 2012

Bardi A (2012) Speciale Global Wind Day – Architettura eolica. http://www.rinnovabili.it/green-building/architettura-eolica39025/. Accessed 2 Oct 2012

Battisti L (2012) Scelta ed installazione delle miniturbine eoliche – Speciale Tecnico QualEnergia.it. http://www.qualenergia.it/sites/default/files/articolo-doc/scelta-installazione-miniturbine-eoliche_speciale_qualenergia_feb2012_0.pdf. Accessed 3 Oct 2012

Buonfrate C (2012) Il settore del minieolico in Italia. http://www.agienergia.it/Notizia.aspx?idd=846&id=58&ante=0. Accessed 2 Oct 2012

Casale C, Viani S, Zanetta GA (2012) Considerazioni tecniche ed economiche sullo stato della tecnologia del minieolico e sul suo sviluppo in Italia. RSE, Milan. http://doc.rse-web.it/?tipo=documento&id=314775. Accessed 23 Nov 2012

Consulente-Energia.com (2011) Minieolico: impianti minieolici verticali e orizzontali. http://www.consulente-energia.com/minieolico-impianto-verticale-vantaggi-prezzi-producibilita-guadagno-mini-eolico-vento-turbine-20-30-40-50-100-150-200-kw-enel-conviene.html. Accessed 2 Oct 2012
D'angelo M (2011) Minieolico: vantaggi e costi di un nuovo impianto. http://minieolico-news.com/minieolico/minieolico-vantaggi-e-costi-di-un-nuovo-impianto. Accessed 2 Oct 2012
Dello Iacovo L (2011) Alta quota e minieolico: l'Italia sulle ali del vento. www.tekneco.it/energia/alta-quota-e-minieolico-litalia-sulle-ali-del-vento/. Accessed 14 Sept 2012
De Nardo L (2010) Eolic Foldable – Un Concept per avere l'energia eolica trasportabile. http://www.100ambiente.it/index.php?/archives/703-Eolic-Foldable-Un-Concept-per-avere-lenergia-eolica-trasportabile.html. Accessed 6 Sept 2012
Elesis Impianti ed Automazioni Srl (2012) WINDSTREAM Micro Turbina Eolica Turbomill 500Wp. http://www.elesisimpianti.it/wordpress/?products=windstream-micro-turbina-eolica-turbomill-500wp. Accessed 14 Sept 2012
ENEA (2009) Il minieolico nel mondo. http://titano.sede.enea.it/Stampa/ skin2col.php?page=eneaperdettagliofigli&id=183. Accessed 15 Feb 2012
Enel.si (2009) Guida agli impianti mini eolici. http://www.google.it/url?sa=t&rct=j&q=guida%20agli%20impianti%20mini%20eolici%20enel.si&source=web&cd=4&sqi=2&ved=0CEcQFjAD&url=http%3A%2F%2Fwww.teatecnologie.it%2Findex.php%3Foption%3Dcom_docman%26task%3Ddoc_download%26gid%3D3%26Itemid%3D&ei=Vxz4UNPOKcjb4QTr0oDwCQ&usg=AFQjCNFx3WogxA9-bAXsLVxqHg-QczwA6w&cad=rja. Accessed 15 Feb 2012
EWEA (2012) Wind in power 2011 European Statistics. http://www.ewea.org/fileadmin/files/library/publications/statistics/Wind_in_power_2011_European_statistics.pdf. Accessed 1 Oct 2012
Facciolla E (2011) Eolico domestico o minieolico: quali sono le offerte sul mercato?. http://www.tuttogreen.it/eolico-domestico-o-minieolico-quali-sono-le-offerte-sul-mercato/. Accessed 29 Sept 2012
Frankl P (2010) La corsa mondiale alle rinnovabili per un futuro low-carbon è già cominciata: l'Italia può e deve parteciparvi. In: ENEA (ed) Le fonti rinnovabili 2010. Ricerca ed innovazione per un futuro low-carbon. ENEA, Rome
Furlan D (2012) Mini eolico ad asse verticale: i produttori. http://www.energie-rinnovabili.net/mini-eolico-ad-asse-verticale-i-produttori. Accessed 26 Sept 2012
Giubilo A (2012) Investire nel minieolico con le nuove tariffe. http://qualenergia.it/articoli/20120829-investire-nel-mini-eolico-con-le-nuove-tariffe. Accessed 2 Oct 2012
GSE (2012) Rapporto Statistico 2011– Impianti a fonti rinnovabili. Accessed 18 Oct 2012
GURI (2012) Attuazione dell'art. 24 del decreto legislativo 3 marzo 2011, n. 28, recante incentivazione della produzione di energia elettrica da impianti a fonti rinnovabili diversi dai fotovoltaici. Official Journal of the Italian Republic no. 159– Ordinary Supplement no. 143. Istituto Poligrafico e Zecca dello Stato, Rome
GWEC (2011) Global Wind Statistics 2011. http://gwec.net/wp-content/uploads/2012/06/GWEC_-_Global_Wind_Statistics_2011.pdf. Accessed 1 Oct 2012
GWEC, Greenpeace (2012) Global Outlook Wind Energy 2012. http://www.gwec.net/wp-content/uploads/2012/11/GWEO_2012_lowRes.pdf. Accessed 30 Nov 2012
ICIM (2011) È di ICIM la prima certificazione dell'eolico in Italia. http://www.icim.it/es/content.604/ita359.html?structureId=5. Accessed 22 March 2012
Legambiente (2012) Comuni rinnovabili 2012. http://issuu.com/legambienteonlus/ docs/rapporto_comuni_rinnovabili_2012/1. Accessed 1 Oct 2012
Microeolico.com (2012) Dove installare un impianto eolico e minieolico. http://www.microeolico.com/dove-installare.html. Accessed 6 Sept 2012
Nonsolosolare (2009) Eolico verticale: Forza 7, la nuova micro turbina tutta italiana. http://www.nonsolosolare.it/2009/12/eolico-verticale-forza-7-la-nuova-micro-turbina-tutta-italiana/. Accessed 15 Feb 2012

Paraggio F (2005) Progetto microeolico – ISF, Ingegneria senza frontiere, in collaboration with ENEA. http://digilander.libero.it/gauguin889/BlogDir/Progetto_ver1_1.pdf. Accessed 10 Oct 2012
Prevailing Energy (2012) Zephyr – turbine eoliche ad asse vertical. http://www.microeolico.it/documentazione/file/scheda_tecnica_ita.pdf. Accessed 2 Oct 2012
Ropatec (2012) Products. http://www.ropatec.com/index_n.php?pid=01&lin=3#. Accessed 10 Oct 2012
SEI—Società Elettrica Italiana (2012) Generatori microeolici minieolici ed eolici. http://www.societaelettricaitaliana.it/impianti_con_generatori_minieolici_microeolici.htm. Accessed 14 Sept 2012
Tita Gallo A (2010) Eolico: la Cleanfield Energy testerà le nuove micro turbine sui tetti di Toronto. http://www.greenbiz.it/panorama/mondo/1321-eolico-la-cleanfield-energy-testera-le-nuove-micro-turbine-sui-tetti-di-toronto-. Accessed 2 Oct 2012
White Wind (2010) Che cos'è il minieolico. http://www.whitewind.it/Home/MiniEolico/tabid/468/Default.aspx. Accessed 16 March 2012

Part III
Experiences and Case Studies in Territorial Contexts

Chapter 17
Territorial Observatory on Sustainability in the Republic of San Marino: A Case Study

Luigi Bruzzi, Simona Verità, Francisco Serrano-Bernardo and Enrique Toscano

Abstract To improve the sustainability of a territory, its development has to be measured through specific indicators. The main function of a Territorial Observatory on Sustainability is to stimulate a positive social attitude towards sustainability by providing the best available information, facilitating the decision-making and stakeholders participation. It can provide various services, such as integrated monitoring of the sustainability of the development, supporting policymaking in prioritization and allocation of resources and community participation process, developing thereby capabilities and knowledge through the supply of information and ensuring appropriate dissemination of results based on scientific research. The chapter describes the process applied to the Republic of San Marino and is based on the results of a study performed in the framework of collaboration between the University of Bologna and the "Coordinamento Agenda 21 San Marino".

Keywords Indicators · Indices · Observatory · Sustainability · Well-being

17.1 Introduction

The local and global environmental situation is in continuous mutation—mainly due to anthropic activities—and ubiquitous: the quality of air, water and other parameters influencing the environmental sustainability (Serrano-Bernardo and Bruzzi 2012), change very rapidly without a clear perception of the general public, demonstrating that monitoring of the environment is imperative (Wursthorn et al. 2011; Wang et al. 2012). For example, the actual situation in the city of Taranto (located in the southern part of Italy and where an important steel manufacturing plant is operating since many years, is supposed to have emitted unspecified quanti-

---

L. Bruzzi (✉)
Department of Physics, University of Bologna, 40100 Bologna, Italy
e-mail: luigi.bruzzi@alice.it

S. Verità · E. Toscano
Bologna, Italy

F. Serrano-Bernardo
Department of Civil Engineering, University of Granada, Granada, Spain

R. Salomone, G. Saija (eds.), *Pathways to Environmental Sustainability*,
DOI 10.1007/978-3-319-03826-1_17,

ties of pollutants) would not have achieved the current dimension if environmental controls, now considered essential, would have been implemented in due course (Cardellicchio et al. 2009; Petronio et al. 2012).

The control of all elements that contribute to worsen the quality of the environment and consequently the health of human beings, is currently recognized as essential to ensure the level of quality of the environment in which we live. The control systems concerning the environmental quality and the level of sustainability require specific tools, that are able to highlight the trends in both, positive and particularly important the negative cases (Moreno-Pires and Fidelis 2012; Shieldsa et al. 2002).

Only in this way, intervention that may allow improving adverse situations can avoid, timely taken, the worsening of the environmental conditions. In order to enable that already existing tools, capable to evaluate the sustainability, can be correctly and well-timed utilized, it is necessary to adopt measures and monitor indicators which consent the quantification of the level of sustainability in a given territory (Pintér et al. 2012). These instruments are usually referred to as sustainability indicators.

17.2 Sustainability Indicators for a Territorial Observatory on Sustainability (TOS)

Sustainability is a process that sets goals that should get closer and closer to the requirements set by the principles of sustainable development. The concept of sustainability can be linked to a given human activity, or a local context, or even a product.

The identification processes of the indicators necessary to identify and measure sustainability are numerous and complex, as well as the system that interconnects them (Floridi et al. 2011). Only one or a limited number of indicators is usually insufficient to determine the level of sustainability. In fact, improving sustainability is a process that is only effectively evaluated if a combination of indicators and their interaction meets the requirements of measuring and monitoring sustainable development in all its complexity (Repettia and Desthieux 2006).

Sustainability indicators implemented through a TOS intend to offer information to local authorities who are committed to achieving a sustainable model of development, and fulfilling at the same time the purpose of providing evidence about their progress to support the prioritization and allocation of resources (OSE 2013). They can be defined, established and monitored through the TOS, which is meant to be a practical tool to verify the evolution of the environmental situation and sustainability for a given territory (Dahl 2012). In this context, the indicators for local sustainability must go beyond the functions typically performed by environmental parameters, overcoming thereby the sectorial approach by taking also into account the economic and social aspects (Gallego Carrera and Mack 2010; Carballo-Penela and Doménech 2010).

This is the reason why "integrated indicators" are defined, which reflects the interplay between environmental, economic and social issues (Bruzzi et al. 2011). One of the important functions performed by the TOS is continuous monitoring of indicators to capture and evaluate their change over a period of time. In this manner, it is possible to identify trends and select intervention directions. The choice of indicators, relevant to characterize a given territory constitutes, hence, the first objective of any proposed TOS (Márton 2012).

Suitable indicators for a TOS should be able to describe demographic data and working activities, physical characteristics of the territory, features of residential areas, types of mobility, production and use of energy, effects on climate related to forest, timber availability, agriculture, hunting and fishing, availability of water, quality of soil and subsoil, air quality, level of noise and radiation, waste management and recycling and presence of polluted sites, are among the most important (OSE 2011; Ministerio de Agricultura, Alimentación y Medio Ambiente 2012).

17.3 Rational for the Creation of a TOS in the Republic of San Marino

The reality of territories is very dissimilar in a variety of aspects. For example, political, economic, social and environmental features can be very different. In order to have a concrete example, in this article the situation of the Republic of San Marino (RSM) is analyzed. Its small territorial size and limited number of inhabitants fits particularly well to the development of an innovative tool such as the TOS, aimed at studying and analyzing the economic, social and environmental sustainability of a small state, following at the same time its development through appropriate monitoring. Furthermore, the establishment of a TOS for the RSM can fulfill all the requirements for a pilot project to be applied to other more extensive situations, through a methodology already tested on a small scale (Scipioni et al. 2009).

17.4 Sustainability Indicators for RSM

The study performed, permitted the selection of the reputed most meaningful indicators and to calculate some of them. The selected indicators were: Gross Domestic Product (GDP), the Human Development Index (HDI), energy, water, municipal waste and air quality indicators, alongside with the carbon and ecological footprint (Maxim 2012).

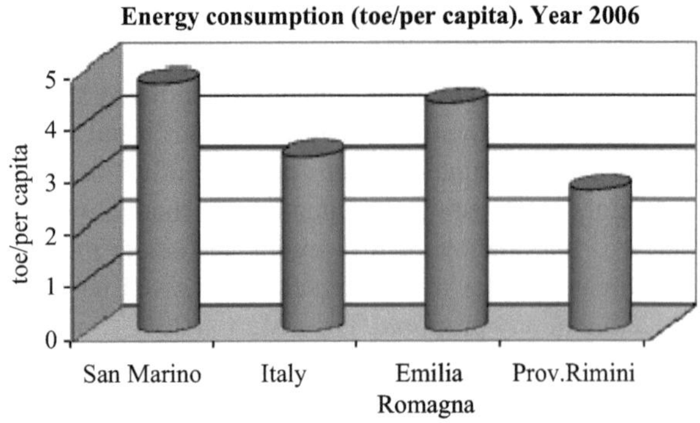

Fig. 17.1 Comparison of the energy consumption per capita of San Marino with Italy, the Italian Region Emilia Romagna and the province of Rimini

17.4.1 Economic and Energy Indicators

The value of GDP for RSM (year 2009) was US$ 1,469 million PPA, corresponding to a GDP per-capita of US$ 47,710 PPA. The comparison with other countries shows an outstanding position of the RSM, second in the rank after Norway and before United States, Ireland and Switzerland (Stratford 1988).

To calculate the HDI, data from the Statistical Office of RSM were used, taking into account the per capita GDP, level of education and life expectancy. The calculated value of HDI amounted to 0.977, locating RSM again in an excellent position, before Norway, Ireland, Netherlands, Sweden and France. For comparison, the HDI for Italy is 0.951 (Morse 2003).

Energy indicators are of paramount importance as they help determine essential parameters such as the dependence on imports, the CO_2 emissions and the amount of emitted air pollutants. Energy management of a territory as small as that of San Marino, without native resources, is dependant mainly on importing from abroad.

Imported supplies include electricity, natural gas—used for heating—and oil derivatives used for transport. For the RSM energy is a vital resource to support the social and economic development. It is therefore important to monitor the parameters that define the resource consumption and its proper use (Streimikiene et al. 2007).

The primary energy annual requirement for the RSM is about 155 ktoe (reference year 2008). The primary energy consumption is characterized by an extensive use of petroleum derivatives and natural gas. This is due to structural components, such as the import of electricity (which constitutes about 38 % of primary energy consumption) and the absence of nuclear generation and renewable sources.

The graph of Fig. 17.1 shows the comparison between the level of consumption of final energy per capita in the RSM to the level in Italy, the Emilia Romagna Region and the Province of Rimini (year reference 2006). The per capita energy consumption of the RSM, is about 30 % greater than the average of Italy, but it is quite close to the average consumption of the Emilia Romagna and significantly higher than the value of the Province of Rimini (+50 %). The differences can be explained by the fact that the RSM has a standard of living relatively high and is considered wealthier as compared to the average in Italy, but similar to the rich region Emilia Romagna. The dissimilarity in the data can be also attributed to the low cost of kWh for the consumers in the RSM as compared to those in Italy. It has also been noted a steady growth in both, primary energy and electricity consumption, that has to be close monitored to establish the subjacent origins, also in view of a desirable improvement of the energy efficiency (Smeets and Weterings 1999).

17.4.2 CO_2 Emissions from the Energy Sector

The RSM produces approximately 350,000 t of CO_2 per year. The data collected over the past 20 years show a growing trend. Italy produces nearly 500 million tons of CO_2 per year, which represents about 2 % of the quantity released globally. This is the expected result of relying on more than 90 % of fossil fuels utilized for the typical end uses of energy: electricity, transport, heating and industry and the increase in the energy consumption per capita. In RSM fuel consumption and thus CO_2 emissions have increased by about 1 % per year, following a general trend and comparable with the consumption of the neighboring territories (Rimini and Emilia Romagna) in Italy.

17.4.3 Indicators for Drinking Water, Air Quality and Municipal Waste

For these indicators only a preliminary analysis has been performed. The indicators analyzed included the consumption of drinking water, the air quality (pollutants, greenhouse gases and particulate matter) and the municipal waste.

The per capita consumption of drinking water in the RSM amounts to about 250 l per day person, a value that is comparable with those of some neighboring municipalities in Italy. The exception being coastal areas, where the average value amount to approximately 300 l/inhabitant per day, with peaks substantially higher in the seasonal period (Gösslinga et al. 2002).

The indicators of air quality, taken as reference, were those that measure the concentration in the air of many pollutants such as oxides of sulfur and nitrogen, particulate matter (PM 10, PM 5, PM 2.5), benzene, Polycyclic Aromatic Hydrocarbons (PAHs), etc. Thematic maps providing information about the winds (speed and predominant directions), temperature, pressure and humidity of the air, were in order to determine the distributions over the territory. The measurements were carried out with mobile and stationary measuring stations.

The preliminary analyses performed in the year 2008, have included the concentrations in air of the PM 2.5, O_3, CO and NO_2. The preliminary results on measurements of other air pollutants demonstrated the ability of the monitoring system to give reliable information on the air quality in significant positions of the RSM territory.

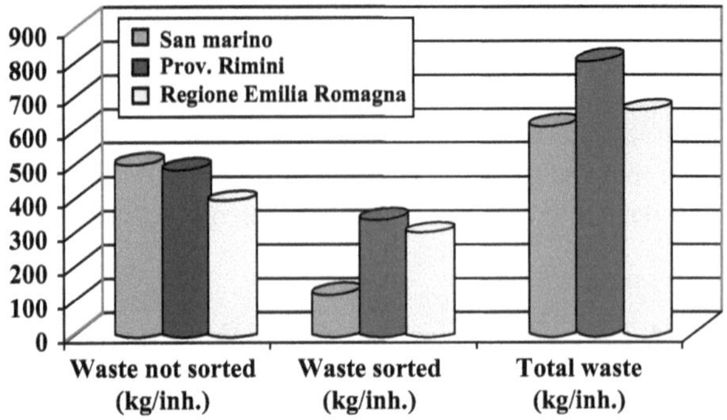

Fig. 17.2 Comparison of the waste production (in kg/inhabitant), in the RSM (*light grey*), the Province of Rimini (*dark grey*) and the Emilia Romagna Region (white)

The production of municipal waste of RSM is approximately 20,000 t per year, equivalent to about 670 kg/inhabitant per year, a value that does not substantially differ much from that of the Emilia Romagna Region, and is much less than the 820 kg/inhabitant of the province Rimini (Fig. 17.2).

On the other hand, the waste sorting amounts to approximately 25 %, relatively low as compared with values of 45.4 % for the region Emilia Romagna and 41.8 % for the Province of Rimini. Thus revealing room for improvement, particularly in the waste sorting, base for transformation of waste in a valuable resource as has been done by other European countries.

17.5 The Ecological Footprint of the Republic of San Marino

The RSM is one of the few countries that have not yet addressed the evaluation of its ecological footprint. Its proper and accurate assessment should be placed as one of the priority actions after the creation of a TOS. At this early stage, due to the limited resources available for the present study, it was not possible to address the calculation of the ecological footprint according to the classical methodology developed by Mathis Wackernagel and William Rees of the University of British Columbia. An alternative approach was adopted, although it does not guarantee the absolute validity of the results has the advantage of raising awareness about the behavioral norms. We opted, therefore, for a demoscopy survey comprising of a questionnaire of simple questions whose answers helped to identify the lifestyle and, henceforth, the sustainability of individual behaviors. This is a methodology extensively used and which has already provided interesting results in many countries.

The use of the information is, however coupled with considerable difficulties in obtaining large numbers of respondents, mainly due to the lack of sensitivity towards sustainability still present in a broad section of the population.

However, there are few but significant examples in which a large numbers of respondents participated in investigations of this type. Among them, there is the foundation ENI Enrico Mattei, which on its site Pandora, (http://www.feem-project.

net/pandora/impronta_eco.php?ids=125) collects the data resulting from the replies of more than 11,000 respondents. Based on these experiences, an opinion survey was carried out for the RSM. The questionnaire selected for the calculation of the carbon and ecological footprint is the one proposed by the WWF (http://www.improntawwf.it/main.php).

Besides the calculation of the ecological footprint, the carbon footprint was also assessed, taking into account that in developed countries the carbon represents more than 50 % of the ecological footprint. This is due to the strong influence of the combustion processes on the level of total sustainability. The survey included different questions to which the respondent had to answer and the replies were analyzed by an appropriate software. This provided the result of the ecological and carbon footprint. The opinion poll involved unfortunately only a low number of people: about fifty families from all over the territory of the Republic making a total of about 160 individuals that provided appropriate answers on the basis of which the calculation was performed. On the same sample, additional processing to understand and better characterize the result was completed.

The average value was 5 ha per capita for the ecological footprint and 12 t of CO_2 equivalent per year for the carbon footprint. For comparison, it should be considered that ecological footprint for Italy is 4.2 ha per capita and that the average of our planet is 1.78 ha per capita.

In order to increase the significance of the results, an investigation on the possible effect of the relative abundance of the members in a family, as well as their gender, was carried out but with no appreciable differences could be established.

17.5.1 Towards Indicators Able to Measure the Well-Being

In the last years close to the concept of sustainability, indicators measuring the well being have started to be developed. These indicators should be able to measure the well being or happiness of a human community, since it is already unanimously recognized that the GDP itself does not well represent the standard of life or the happiness in a society. The first step forward to overcome this situation was the introduction of the already discussed HDI. Recently, studies and analyses have been developed, attempting to express the level of wellness of a given society through a larger number of parameters, than those included in the calculation of HDI. This possibility has been the subject of a study by the Commission for the Measurement of Economic Performance and Social Progress, instituted in 2008. The report produced by the this so-called Stiglitz Commission, proposes to introduce more parameters, including participation in politics, social relationships, environmental condition and physical safety, being among others. These indications should converge to calculate an index of well being, be able to express and measure the quality of life (Stiglitz et al. 2009).

The proposal to introduce an indicator related not only to the economy, as it is the case for the GDP, is not new. The need of a novel approach to guide the development was premonitory stated in 1968 by Robert Kennedy, who prophetically warned:

> Yet the gross national product does not allow (to evaluate) for the health of our children, the quality of their education, or the joy of their play. It does not include the beauty of our poetry or the strength of our marriages; the intelligence of our public debate or the integrity of our public officials.

A TOS in the RSM should also consider this aspect, although the high standard of life of the country anticipates an excellent result also for this type of index. During the last two years the studies to introduce indicators that are able to measure the level of well-being have continued. An important initiative was taken from two Italian important public institutions: ISTAT (Istituto Nazionale di Statistica) and CNEL (Consiglio Nazionale dell'Economia e del Lavoro). Starting with the 8 indicators the study proposes to introduce 4 more to make a total of 12 (the 12 dimensions of well-being: environment, health, economic well-being, education, working activities, social relationships, safety, subjective well-being, landscape and cultural heritage, research and innovation, quality of services, politics and innovation. Each one of these dimensions constitutes a domain in which an average of 11 high quality statistical indicators are included thus making a total of 134. This advanced approach can in principle take into account a lot of different conditions including: percentage of burglary, availability of green park areas included in urban areas, CO_2 emissions, average income per family, physical and psychological status of individuals (Kaneda et al. 2011).

17.6 Conclusions

In any given state, even when goals have been agreed and set, policy-makers need to prioritize and allocate resources in the face of eventual conflicting requirements concerning economic growth and social progress in conjunction with environmental sustainability. Wrong decisions can have serious consequences, e.g., causing a crisis or even leading to an increase of human suffering. Improving the basis for sound decision-making, integrating many complex issues in comprehensible indexes, providing simple signals to the decision makers are thus of paramount importance. In the present time, when modern technologies increase the flow of information but not necessarily our ability to properly evaluate it, we need tools that condense information for rapid assimilation, rising thereby awareness.

This is the key role of a TOS, based on the adoption, measurement and monitoring of suitable indicators. In general it can be safely stated that there is always room for improvements in any given anthropic activity. The present preliminary assessment of important sustainability indicators towards the establishment of a TOS in the RSM shows that some of them display potential for development. They also provided the basis, and have been used, for international comparisons in support of the pursuing of sustainability goals and targets.

The country can now design and implement a national program for sustainability, based on data that go far beyond traditional economic and social statistics. This type of collecting and evaluation of data campaign can be adapted and customized to similar realities in comparable territories. In this context it has to be kept in mind that the priority is to increase the use of information for policy making, planning and management of resources in a sustainable manner. According to present international experience, defining indicators and using them to support decision-making has proven to be most cost-effective towards sustainable development, under the general principle that prevention is always better than cure.

Acknowledgments The authors would like to thank Emanuele Guidi and Antonio Conti of the "Coordinamento Agenda 21" of San Marino for supporting the study by collecting data and develop the analysis contained in this chapter.

References

Bruzzi L, Boragno V, Serrano-Bernardo F, Verità S, Rosúa-Campos JL (2011) Environmental management policy in a coastal tourism municipality: the case study of Cervia (Italy). Local Environ 16(2):93–113. http://dx.doi.org/10.1080/13549839.2011.558075

Carballo-Penela A, Doménech JL (2010) Managing the carbon footprints of products: the contribution of the method composed of financial statements (MC3). Int J Life Cycle Assess 15:962–969. doi:10.1007/s11367-010-0230-1

Cardellicchio N, Buccolieri A, Leo AD, Librando V, Minniti Z, Spada L (2009) Methodological approach for metal pollution evaluation in sediments collected from the Taranto Gulf. Toxicol Environ Chem 91(7):1273–1290

Dahl AL (2012) Achievements and gaps in indicators for sustainability. Ecol Indic 17:14–19. doi:10.1016/j.ecolind.2011.04.032

Floridi M, Pagni S, Falorni S, Luzzati T (2011) An exercise in composite indicators construction: assessing the sustainability of Italian regions. Ecol Econ 70:1440–1447. doi:10.1016/j.ecolecon.2011.03.003

Gallego Carrera D, Mack A (2010) Sustainability assessment of energy technologies via social indicators: results of a survey among European energy experts. Energy Policy 38:1030–1039. doi:10.1016/j.enpol.2009.10.055

Gösslinga S, Hanssonb CB, Hörstmeierc O, Saggeld S (2002) Ecological footprint analysis as a tool to assess tourism sustainability. Ecol Econ 43:199–211. doi:10.1016/S0921-8009

Kaneda T, Lee M, Pollard K (2011) SCL/PRB index of well-being in older populations final report, Global Aging and Monitoring Project. http://longevity3.stanford.edu/wpcontent/uploads/2011/09/SCL-PRB-Index-of-Well-Being-in-Older-Populations.pdf. Accessed 15 Nov 2012

Márton P (2012) A territorial understanding of sustainability in public development. Environ Impact Assess Rev 32:61–73. doi:10.1016/j.eiar.2011.03.004

Maxim M (2012) Building shared socio-economic indicators for biodiversity. A case study in the Ile-de-France region (France). Ecol Indic 13:347–357. doi:10.1016/j.ecolind.2011.05.015

Ministerio de Agricultura, Alimentación y Medio Ambiente (2012) Sostenibilidad en España 2012. Secretaría General Técnica. Centro de Publicaciones, Madrid

Moreno-Pires S, Fidelis T (2012) A proposal to explore the role of sustainability indicators in local governance contexts: The case of Palmela, Portugal. Ecol Indic 23:608–615. doi:10.1016/j.ecolind.2012.05.003

Morse S (2003) For better or for worse, till the human development index do us part? Ecol Econ 45:281–296. doi:10.1016/S0921-8009

OSE (Observatorio de la Sostenibilidad en España) (2011) Sustainability indicators. http://www.sostenibilidad-es.org/es/indicadores?field_anyo_value_many_to_one=2011. Accessed 22 Feb 2013

OSE (Observatorio de la Sostenibilidad en España) (2013) Memoria de Actividades 2011. http://www.sostenibilidad-es.org/sites/default/files/_Documentos/memoria_actividades_ose_2011_pre_def_peque.pdf. Accessed 22 Feb 2013

Petronio B, Cardellicchio N, Calace N, Pietroletti M, Pietrantonio M, Caliandro L (2012) Spatial and temporal heavy metal concentration (Cu, Pb, Zn, Hg, Fe, Mn, Hg) in sediments of the Mar Piccolo in Taranto (Ionian Sea, Italy). Water Air Soil Pollut 223:863–875

Pintér L, Hardi P, Martinuzzi A, Hall J (2012) Bellagio STAMP: principles for sustainability assessment and measurement. Ecol Indic 17:20–28. doi:10.1016/j.ecolind.2011.07.001

Repettia A, Desthieux G (2006) A relational indicator set model for urban land-use planning and management: methodological approach and application in two case studies. Landscape Urban Plan 77:196–215. doi:10.1016/j.landurbplan.2005.02.006

Scipioni A, Mazzi A, Mason M, Manzardo A (2009) The dashboard of sustainability to measure the local urban sustainable development: the case study of Padua Municipality. Ecol Indic 9:364–380. doi:10.1016/j.ecolind.2008.05.002

Serrano-Bernardo F, Bruzzi L (2012) Introducción al estado del ambiente. In: Serrano-Bernardo F, Bruzzi L (eds) Gestión Sostenible del Ambiente: Principios, contexto y métodos. Editorial Universidad de Granada, Granada

Shieldsa D, Šolarb SV, Martinc WE (2002) The role of values and objectives in communicating indicators of sustainability. Ecol Indic 2:149–160. doi:10.1016/S1470-160X

Smeets E, Weterings R (1999) Environmental indicators: typology and overview. http://www.eea.europa.eu/publications/TEC25. Accessed 15 Nov 2012

Stiglitz E, Amartya S, Fitoussi JP (2009) Report by the Commission on the Measurement of Economic Performance and Social Progress. http://www.stiglitz-sen-fitoussi.fr/documents/rapport_anglais.pdf. Accessed 15 Nov 2012

Stratford JS (1988) United States economic indicators: definitions, sources, and resources. Gov Publ Rev 15:231–236. doi:10.1016/0277-9390

Streimikiene D, Ciegis R, Grundey D (2007) Energy indicators for sustainable development in Baltic States. Renew Sustain Energy Rev 11:877–893. doi:10.1016/j.rser.2005.06.004

Wang H, Bai H, Liu J, Xu H (2012) Measurement indicators and an evaluation approach for assessing Strategic Environmental Assessment effectiveness. Ecol Indic 23:413–420. doi:10.1016/j.ecolind.2012.04.021

Wursthorn S, Poganietz WR, Schebek L (2011) Economic-environmental monitoring indicators for European countries: a disaggregated sector-based approach for monitoring eco-efficiency. Ecol Econ 70:487–496

Chapter 18
The European Policy for the Sustainability of Urban Areas and the "Covenant of Mayors" Initiative: a Case Study

Mariarosaria Lombardi, Roberto Rana, Pasquale Pazienza and Caterina Tricase

Abstract The "Covenant of Mayors" (CoM) is an initiative of the DG for Energy of the European Commission. It was started in 2008, with the aim of reducing Greenhouse Gases (GHGs) by 20 % by 2020 through the implementation of adequate actions taken in agreement between the European Commission and the municipalities (so called signatories), which intend to adhere. The actions are identified and listed in the "Sustainable Energy Action Plan (SEAP)," which each signatory must produce within one year from adhesion and submit to the European organisms for formal approval and consequent accession to financial tools to implement them. A total of 36 (out of 61) municipalities of the province of Foggia subscribed to the above-mentioned agreement in 2010 and, 1 year later, submitted their SEAPs to the validation of the experts of the Joint Research Centre-Institute of Energy (JRC-IE) of the European Commission. All the submitted SEAPs were formally approved in July 2012. This chapter reports on the methodology the authors used for the elaboration of the SEAPs and the achieved results.

Keywords Covenant of Mayors · Energy policy · Greenhouse gases · Sustainable energy action plan · Urban planning

18.1 Introduction

The "CoM" is an initiative of the DG Energy of the European Commission started in 2008 and dedicated to the municipalities of European and extra-European countries which intend to reduce and/or contain by 2020 the energy consumption associated to the use of fossil fuels (and the GHGs as a result) by 20 %. For this reason, the "Covenant of Mayors" initiative can be surely seen as a very important tool for the actual implementation of the European climate-energy policy also called the policy of 20-20-20 by 2020. In fact, municipalities can significantly contribute to achieving the above targets and to reducing the CO_2 by 13 % in those activity sectors

C. Tricase (✉) · M. Lombardi · R. Rana · P. Pazienza
Department of Economics, University of Foggia,
L.go Papa Giovanni Paolo II n. 1, 71121 Foggia, Italy
e-mail: caterina.tricase@unifg.it

R. Salomone, G. Saija (eds.), *Pathways to Environmental Sustainability*,
DOI 10.1007/978-3-319-03826-1_18,

which are not considered in the EU Emission Trading System (i.e. civil, residential, transport, small and medium enterprises), since they can exert a major level of control in agreement with the Decision of the EU Commission no. 406/2009, also called "Effort Sharing" (Lumicisi 2012). The "CoM" initiative implies that the municipality taking part must produce and submit to the European Union—within one year from the signature of the agreement with the DG for Energy—a SEAP to reduce GHG emissions. This planning document must then be submitted for validation to the attention of the experts of the Joint Research Centre-Institute of Energy (JRC-IE) of the European Commission, which—in Italy—is based at Ispra. The validation allows the participating municipality to access financial tools for the actual implementation of its SEAP. This document is basically composed of two different sections: the Baseline Emission Inventory (BEI) and the strategic action plan. The first section is a typical energy balance in which the energy production and consumption in the considered municipality are estimated in terms of Megawatt per hour (MWh) together with the resulting GHGs (in particular CO_2 that represents the main greenhouse gas released by urban areas). This is required to be done while considering emission generating activity sectors (public, civil, tertiary, transport, and industrial) and the energy vector (electricity, natural gas, LPG, gasoline, diesel, wood, etc.). The second section describes the actions (particularly referred to sustainable mobility, public building energy efficiency, installation of renewable energy production plants, etc.) through which to decrease the GHGs computed in the BEI (Lombardi et al. 2011). In November 2012, about 4,500 municipalities (accounting for about 34 % of the total European population, that is more than 170 million—out of 502 million—EU inhabitants) had signed the adhesion to the "Covenant" initiative. The member States most involved in this initiative are: Italy (with about 51 % of the municipalities) and Spain (26 %). They are followed by France (3.2 %), Romania (2.3 %), Greece (2.2 %), Portugal (1.7 %) etc. So far, 1,940 SEAPs have been submitted for validation, of which 482 have already been accepted and 36 have been refused. The key role of Italy in the implementation of the "CoM" initiative is clear and the number of participating municipalities will further increase in the near future. According to official Italian figures, 748 signatories submitted their SEAPs and, so far, 279 have received validation. All this has been possible also thanks to the role played by the activation of "supporting structures," such as groups of municipalities, provincial and regional institutions (EuMayors 2012). In Southern Italy a relevant action was started in 2010 by 36 (out of 61 total) municipalities in the province of Foggia. All these—under the coordination of the Province of Foggia—signed the "Covenant" agreement with the DG for Energy in 2010 and in May 2011 submitted their respective SEAPs, which were validated and accepted by the JRC in July 2012. The production of the SEAPs was really complex and difficult due to the lack of information associated to the energy consumption and production at municipal level. The aim of this chapter is to briefly report on the methodology which was used for the SEAPs production and how this contributes to the realization of the environmental sustainability of the territory of the province of Foggia.

18.2 Materials and Methods

18.2.1 Baseline Emission Inventory

The production of BEI is principally based on the methodology proposed in the guidelines of the JRC of the European Union (Bertoldi et al. 2010). According to these rules it has been identified a specific year starting from which a complete and detailed set of energy consumption data could be gathered. It was chosen the year 2005 (closest to 1990—the year suggested by the EC—for which data can be collected as complete and reliable as possible) as the reference for the calculation of the needed values which only took into consideration CO_2 emission, and not all the GHGs, with the aim of implementing a strategy of analysis more standardized and broadly accepted. In fact worldwide 80 % of GHGs are released in urban areas and 90 % of these are attributable to carbon dioxide because the result of energy combustion in residential, transport and tertiary sector (IEA 2012; Dickinson and Tenorio 2011). For a better specification, the CO_2 emissions were computed while considering both direct (namely those deriving from fuels consumption and from electricity production within the municipalities borders) and indirect emissions (specifically those generated outside the municipalities' borders and deriving from the production of imported electricity).

This was made in accordance to what the IPPC suggests as highlighted by since it guarantees a standardized and simpler procedure (Lombardi et al. 2011; Yu et al. 2012). In addition, the analysis focused on both public and private sectors and always following the JRC guidelines. This main categories were considered: 1) buildings, tools/plants and industries; 2) transport.

For the *public* sector, a questionnaire was sent to all municipalities which had adhered to the "Covenant of Mayors" in which they were asked to give specific information about buildings (structural characteristics, energy consumption, the use of Renewable Energy Source (RES) installations), lighting, vehicle fleets (type of vehicles, year of matriculation, fuel consumption, etc.) and other initiatives in the environmental sector.

For the *agricultural, tertiary, non-ETS industry, urban transport and residential* sector, as the authors were unable to use a questionnaire because of the difficulties of data retrieval, they proceeded in estimating the energy consumption for each sector and vector. Although this could generate uncertainty in evaluating the investigated phenomena, it achieved plausible information with respect to the aims it had reach. For example, the quantity of energy, needed for heating and transport in the agricultural and non-ETS industries, was calculated by dividing the provincial consumption for each single fuel (natural gas, diesel, gasoline and LPG)—supplied by the Financial Technical Office of Foggia (Ufficio Tecnico di Finanza di Foggia (UTF) 2010, personal communication)—by the total number of workers in each sector (IPRES 2006). The average value obtained was then multiplied by the total number of those employed in each municipality. Regarding

Table 18.1 Year energy consumption per capita by economic sectors

	Agriculture	Residential	Tertiary	Non ETS Industry[c]
LPG (l/capita)	None	28.34	245.80	None
Natural gas (m^3/capita)[a]	None	338[b]	1,000	6,523
Gasoline (l/capita)	482	–	–	–
Diesel (l/capita)	11,353	3.95	17.97	1,724

[a] Those municipalities which were not supplied with natural gas in 2005 have been excluded from the calculation

[b] This is the average value per resident reported for the city of Foggia by ISTAT in 2005. This parameter was preferred instead of the estimation deriving from the calculations done for diesel and LPG because the result would have been too low and unrealistic

[c] ETS - Emission Trading System

the agricultural sector it is evident that should be considered other GHG emissions (CH_4 and N_2O) apart from CO_2. Nerveless, since this sector is not included in the action strategy for emissions abatement it has focused on CO_2 only, without considering the other gases.

In the case of the energy needed for the heating of residential and tertiary buildings, the provincial data for each single vector (LPG, diesel and natural gas)—sourced by the UTF—was reported in terms of the sum of both. Therefore, it was necessary to split this sum according to the Italian average of percentage associated to the considered type of buildings as identified by the ENEA (ENEA 2009). The obtained values were then divided by the total provincial population (residential) and by the number of workers in the tertiary sector as of 2005. The result was then multiplied by the number of inhabitants of the respective municipalities (see Table 18.1). For domestic heating, the use of forest biomass was also considered to take into account some behavioral features of some local communities. To this purpose, the information on consumption was deduced from the analysis of the authorization requests for forest cut given by the Regional Office for forest management.

The consumption of electrical energy by private individuals was deduced by the IPRES (2006) evaluation.

For urban transport, when local city data regarding the use of fuel (diesel oil, LPG, gasoline and natural gas) was unavailable from UTF, the number of different types of vehicles present in the urban area (motorcycles, cars, articulated vehicles, etc.) (Automobile Club d'Italia (ACI) 2010, personal communication) was multiplied by the average consumption and by the average yearly number of kilometers travelled (ISPRA/APAT 2010).

Once the fuel and electricity consumption per sector and energy vector were obtained, they were transformed into MWh and CO_2. The European Union guidelines for conversion factors were referred to with regard to MWh, while for carbon dioxide those of the IPCC (Intergovernmental Panel on Climate Change) were used which take into consideration all CO_2eq and CO_2 emissions deriving from fuel use (Table 18.2) (IPCC 2006).

Table 18.2 Some energy conversion factors utilized in the considered SEAPs

Conversion factors	Natural Gas	LPG	Diesel	Gasoline	Electr.	Wood
Specific weight (kg/m³)* (kg/l)	0.800	0.600	0.832	0.734	–	–
t→MWh	13.300	13.100	11.900	12.300	–	4.300
MWh→tCO_2	0.202	0.231	0.267	0.249	0.483	–
l→kWh	–	7.600	10.000	9.200	–	–
t→tep	–	1.100	1.080	1.020	–	0.371
tep→MWh	–	–	–	–	11.600	–

* only for natural gas

18.2.2 Action plan

In compiling the Action Plan, the authors only considered the CO_2 emissions which could be reduced as a result of the implementation of interventions and policies for which the municipal governments are typically responsible. In particular, reference was made completely to the public, residential and tertiary sectors and partially to non-ETS industry (heating/cooling systems, lighting and electrical motor consumption) and transport (mobility in the urban area). Agriculture, some forms of transport (out-of-town transport) and industry (consumptions deriving from productive processes) were not counted (Città di Avigliana 2010). The objective of the Action Plan was established to be the overall reduction of 21 % of the emission of carbon dioxide, with respect to the values calculated in 2005, and that the public sector would reduce its emissions by 50 %. This comes from its educational role towards its citizens and businesses and from the direct control that the administration holds over the actions anticipated by the Plan. With this aim, improvements to building insulation in conjunction with extra maintenance work, the replacement of public lighting, the purchase of ecological vehicles or of energy-saving tyres were planned. In the transport sector, a percentage of 10 % was established in encouraging citizens to change their transport habits within the local council area.

The Energy Action Plan is composed of 27 Actions divided into 6 sectors:

1. Information/Training (INFO)—INFO 01 Communication campaign for sustainable mobility, INFO 02 Covenant of Mayors online, INFO 03 Energy guardian, INFO 04 Energy desk, INFO 05 Agency for inter-council energy, INFO 06 Training sessions, INFO 07 Refresher courses for the building sector, INFO 08 School training days;
2. Public (PA)—PA 01 Urban forestation, PA 02 Green public procurement, PA 03 Regulation for building energy efficiency, PA 04 Implementation of building energy efficiency, PA 05 Energy production from renewable sources, PA 06 Public lighting, PA 07 Renewable energy purchase from private companies;
3. Residential (RES)—RES 01 Energy-saving incentives;
4. Tertiary (TER)—TER 01 Zero km Food products, TER 02 UNI-EN ISO 14001 or 16000—EMAS, TER 03 Electrical engine repowering;

5. Industrial (IND)—IND 01 UNI-EN ISO 14001—EMAS, IND 02 Electrical engine repowering; and
6. Mobility (MOB)—MOB 01 Substitution of municipality vehicle fleet; MOB 02 Substitution of tyres with ecological ones; MOB 03 Improvement of ICT services; MOB 04 Municipality wireless networks; MOB 05 Bike sharing service; MOB 06 Creation of cycle and pedestrian routes.

For each of the above-mentioned actions, an initial analysis of their economic feasibility in terms of pay-back time was carried out. At the same time, the following issues were also dealt with: a brief description of the technical actions to be implemented together with the individuals (private and public) to be involved, the expected results both in terms of economic, environmental (tCO_2/year) and energy (MWh/year) savings.

18.3 Results and Discussion

As has already been said, the production of the SEAPs referred to 36 municipalities with a combined population of over 260,000 people (out of 640,836 people in the whole province) distributed over nearly 3,400 km^2 (out of a total provincial area of 6,971 km^2). With the exception of Cerignola (58,827), Monte Sant'Angelo (13,250), Orta Nova (17,767), San Marco in Lamis (14,576), San Severo (55,399) and Torremaggiore (17,365), which are the most populated municipalities, the majority of the municipalities are small in size (on an average no more than 2,700 inhabitants) and are mainly concentrated in the hill and mountain areas of the province of Foggia.

As a result of the implementation of the actions listed in the SEAPs and as can be seen in Table 18.3, the yearly saving of carbon dioxide is equal to 136,195 t. The section of action which contributes the most to the reduction of the considered emissions is that of "INFO," although its contribution is difficult to actually quantify, not having available standardized procedures. In fact, it operates through propaedeutic actions which do not exert any direct impact on the emission decrease but it works mainly to increase the sensitivity and knowledge of the theme of energy efficiency, the use of renewable sources etc.. Even the section "OTHER ACTIONS" works to help the reduction of CO_2. In particular, action PA07—the purchase of wind energy by private companies—appears to be one of the most important factors. The authors believe that in those areas where wind energy plants have been installed, it is possible to purchase renewable energy which has not been put into the national grid (since the national grid is not able to cope with all the energy produced) by signing ad hoc agreements with the energy producers. This could give the opportunity of retrieving energy produced at advantageous prices with consequent environmental savings. With regard to this, we have cautiously estimated a 10 % reduction in consumption of energy from fossil fuel sources and a saving of carbon dioxide equal to 5,630 t/year (27 % of the total). Action PA03, regarding the adoption of a Regulation for building energy efficiency, aims to promote the construction of buildings with more superior energy and environmental saving features according to what is established

Table 18.3 Type of interventions planned in the SEAPs of the municipalities of the province of Foggia and carbon dioxide saved (tCO_2)

Municipality	EE[a]	PL[b]	RES[c]	MOB[d]	INFO[e]	Other actions[f]
Accadia	29	121	0	10	679	432
Alberona	92	0	0	11	191	110
Biccari	24	114	31	18	1,331	214
Carapelle	43	127	46	126	1,873	106
Carlantino	9	0	15	9	186	25
Casalnuovo Monterotaro	12	62	50	19	213	35
Casalvecchio di Puglia	0	0	83	20	76	85
Castelluccio Valfortore	24	0	9	21	332	25
Castelnuovo della Daunia	20	0	25	13	233	503
Celenza Valfortore	43	12	36	24	498	129
Celle di San Vito	37	29	101	1	35	25
Cerignola	353	886	272	1,393	21,995	3,530
Deliceto	581	189	53	30	1,623	726
Faeto	52	91	0	18	74	90
Lesina	41	0	49	163	3,100	493
Mattinata	103	214	83	74	2,787	402
Monte Sant'Angelo	97	0	19	166	6,019	1,522
Monteleone di Puglia	66	0	0	15	246	156
Motta Montecorvino	28	75	0	19	172	215
Ordona	22	99	76	21	912	109
Orsara di Puglia	28	124	0	50	901	319
Orta Nova	127	0	76	318	6,863	759
Pietramontecorvino	44	162	77	19	631	438
Rocchetta Sant'Antonio	27	0	64	8	327	396
Roseto Valfortore	39	0	30	7	149	123
San Marco in Lamis	315	378	68	149	5,092	846
San Marco la Catola	0	0	30	8	261	21
San Paolo di Civitate	55	235	26	347	2,217	373
San Severo	516	1,679	318	4,562	26,399	5,098
Sant'Agata di Puglia	71	12	0	15	194	644
Stornara	77	204	76	39	1,596	248
Stornarella	0	237	0	97	1,601	291
Torremaggiore	0	559	0	159	5,360	964
Troia	60	214	0	47	1,793	1,215
Volturara Appula	26	13	0	13	94	72
Volturino	14	152	76	25	205	305
Total	3,075	5,990	1,790	8,033	96,260	21,047

[a] EE: Energy Efficiency and saving to which public building restructuring (PA04)
[b] PL: substitution of Public Lighting systems with new energy saving technology (PA06)
[c] RES: use of Renewable Energies Source like photovoltaic systems, micro wind turbines, and teleheating (PA05)
[d] MOB: replacement of vehicles and tires (MOB01 and MOB02), strengthening of telematic service (MOB03), development of local wireless web (MOB04), bike sharing service (MOB05) and construction of cycle and/or pedestrian route (MOB06)
[e] INFO: information/communication on energy saving activities of municipality
[f] OTHER ACTIONS: in this column different actions are collected such as incentives for energy efficiency (RES), urban forestation (PA01), regulation for building energy efficiency, (PA03), purchase of renewable energy from private companies (PA07) action on TERtiary (TER) and action on INDustrial sector (IND)

by more recent European and national laws. This action results in a large reduction of CO_2, with an environmental saving of about 6,228 tCO_2/year (about 30 % of the total sector). The emissions in the "MOB" section are quantitatively less relevant because they derive mainly from the increased request for the substitution of the old municipality vehicles with newer and more environmentally-friendly ones. The greatest contribution was given by the implementation of a bike sharing service (in the municipalities of San Severo and San Paolo di Civitate) and the construction of a cycle-pedestrian route of 1.5 km in the municipality of Carapelle. These actions brought about a 50 % reduction in carbon dioxide emissions in the entire section.

The "PL" section mainly refers to the substitution of mercury light bulbs with high pressure sodium light bulbs, also including the implementation of light flow regulators. We thought to avoid installing LED lighting due to the high cost of their purchase and maintenance. The resulting environmental saving is equal to about 6,000 t/year. The "EE" mainly refers to the renovation and/or substitution of old parts, such as the external insulation of a house (walls and roof insulation), the installation of new and energy-efficient windows and doors, the installation of new and energy-efficient heating systems (e.g., implementation of thermal convection or application of thermostatic valves), thermal regulation in buildings, etc.. Out of a total of 489 buildings, 142 are those to be subject to these kind of interventions. The resulting reduction of carbon dioxide levels is estimated to be more than 3,000 t/year. The actions referring to the implementation of renewable energy plants, regards the installation of wind and photovoltaic micro-plants. In addition, the installation of three biomass cogeneration power plants was considered for the municipalities of Celle di San Vito, Orsara di Puglia and Pietramontecorvino. The environmental saving deriving from these type of actions was computed to about 200 t/year.

The following graph (Fig. 18.1) highlights the percentage of CO_2 abatement—or, in other terms, the environmental saving—for each single municipality as the result of the implementation of the actions identified in the SEAPs. It is evident how the majority of the municipalities overtake the minimum target of reducing the GHG emissions which is fixed at 21 %. The most impressive case is that of Celle San Vito, whose environmental saving stands at 72 % with respect to the situation observed at 2005. This situation is the result of the planned implementation of a cogeneration plant fed with biomass for teleheating and whose power is sufficient to guarantee the provision of heat and renewable energy for the whole local population (about 150 inhabitants).

18.4 Conclusions

The aim of the "CoM" initiative is to promote the active participation of municipalities and their populations in the identification and implementation of sustainable development models at the local scale. In other words, the "CoM" initiative intends to achieve significant levels in the delocalization of energy and environ-

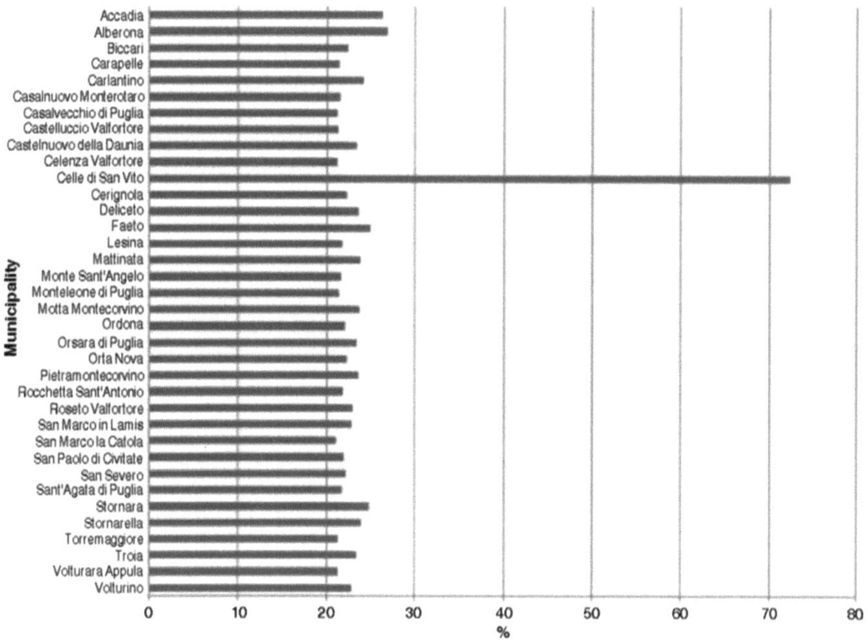

Fig. 18.1 Percentage of CO_2 abatement for each single municipality as the result of the implementation of the actions identified in the SEAPs

mental policy decisions. In this context, the experience matured by the province of Foggia (Italy)—a territory in Southern Europe where the environmental sustainability management of the urban areas is still far away from a concrete and effective execution—is a virtuous example of a strong commitment towards the implementation of the environmental and energy guidelines of the European Commission. As highlighted in the chapter, various technical difficulties were met while producing the SEAPs. Most of these difficulties were associated to the lack of adequate data per activity sector and energy vector at municipal level. In addition, one of the main features of the majority of the considered municipalities is the small size and low level of industrialization. For this reason, it was often necessary to re-adapt and modify the indications supplied by the JRC-IE. Nevertheless, the SEAPs were produced and submitted within the given deadlines and have all been accepted by the European Commission, and they are now waiting for the financial resources to implement the actions planned. With regard to the financial aspect, a final consideration should be made while stressing the necessity to align the implementation of the SEAPs with regional funding. At the time of writing, in fact, there is still a wide gap between the European policy indication associated to the implementation of the "Covenant" initiative and the way in which—especially the regions categorized as convergence areas—have planned to use the funds they receive from the European Union. Solving this problem could represent an accelerator in the implementation of the SEAPs and one more step in the right direction to pursue the energy and environmental sustainability of the considered local territories.

References

Bertoldi P, Bornàs Cayuela D, Monni S, de Raveschoot RP (2010) Guidebook: "how to develop a sustainable energy action plan (SEAP)." Joint Research Centre Scientific and Technical Reports, Publications Office of the European Union, Luxembourg. http://publications.jrc.ec.europa.eu/repository/bitstream/111111111/14204/1/com%20guidebook%20jrc%20format.pdf. Accessed 25 Oct 2012

Città di Avigliana (2010) Piano di Azione sull'energia - Avigliana in transizione, aprile 2010. http://www.comune.avigliana.to.it. Accessed 18 Oct 2012

Dickinson J, Tenorio A (2011) Inventory of New York City greenhouse gas emissions. Mayor's Office of Long-Term Planning and Sustainability, New York. http://nytelecom.vo.llnwd.net/o15/agencies/planyc2030/pdf/greenhousegas_2011.pdf. Accessed 31 Jan 2012

ENEA (2009) Rapporto energia e ambiente 2007-2008 I dati - Statistiche energetiche e ambientali - Distribuzione dei consumi energetici del civile tra terziario e residenziale per fonte energetica. http://www.enea.it. Accessed 1 Nov 2012

EuMayors.eu (2012) Covenant in figure. http://www.covenantofmayors.eu/about/covenant-in-figures_en.html. Accessed 14 Nov 2012

Intergovernmental Panel on Climate Change (IPCC) (2006) 2006 IPCC Guidelines for national greenhouse gas inventories. In: Eggleston S, Buendia L, Miwa K, Ngara T, Tanabe K (eds) Institute for Global Environmental Strategies (IGES), Hayama, Japan. http://www.ipcc-nggip.iges.or.jp. Accessed 3 Nov 2012

International Energy Agency (IEA) (2012) World Energy Outlook 2011. http://www.worldenergyoutlook.org. Accessed 31 Jan 2012

ISPRA/APAT (2010) Capitolo 3- Trasporti, in Annuario dei dati ambientali 2009. Edizioni Ispra Roma. http://www.isprambiente.gov.it/files/pubblicazioni/rapporti/annuario-2009/3-trasporti.pdf. Accessed 11 Oct 2012

Istituto Pugliese di Ricerche Economiche e Sociali (IPRES) (2006) La Puglia in cifre. In: Ruzzo M . Sedit srl, Bari

Lombardi M, Rana R, Tricase C (2011) Per la sostenibilità ambientale delle città. In: Campisi B, Novelli V Contribution of Commodity Science to a Sustainable World - Proceedings of the 25th Italian Congress of Commodity Science, FORUM, Edizioni, Udine (Italy)

Lumicisi A (2012) Il Patto dei Sindaci: un progetto di sviluppo sostenibile per le città d'Europa. Proceedings of the conference "Il Patto dei Sindaci: uno strumento per la città sostenibile," Soveria Mannelli (Italy), 12 June 2012. http://www.energiacalabria.org/lumicisi.pdf. Accessed 1 Nov 2012

Yu W, Pagani R, Huang L (2012) CO_2 emission inventories for Chinese cities in highly urbanized areas compared with European cities. Energy Policy 47:298–308

Chapter 19
Tools for Promoting Environmental Sustainability in Poland

Joanna Kulczycka and Lukasz Lelek

Abstract There are many EU documents which promote sustainable development, e.g. EU Sustainable Development Strategy, Europe 2020 Flagship Initiative-Innovation Union, EC Communication on a Roadmap to a Resource-Efficient Europe, as well as the Eco-innovation Action Plan. One of the important features of these documents is their continuous promotion of the development of eco-innovation and the implementation of environmentally-friendly technologies. They can play a key role in raising standards of living contributing to decrease material inputs, reduce energy consumption and emissions, recover valuable by-products and minimize waste. There are many different tools and incentives for the development and assessment of pro-ecological solutions i.e. Eco-Label, ETAP, SCP/SIP, EMAS, IPPC, Energy-using Product Directive, GPP. In the EU and also in Poland there are policies for increasing environmental standards and decreasing the negative impact on the environment by using both financial and non-financial incentives. One of the most popular tools is financial support for investment projects contributing to sustainable development, which is available from EU funds (Structural Funds) and from different state funds at both national and regional levels. The aim of this chapter is to analyze the indicators for evaluating environmental effects used by different environmental funds in Poland and to propose Life Cycle Assessment (LCA) as a tool for assessment of the environmental impact of investment projects supported by these funds.

Keywords Investment project · LCA · Municipal waste management · Incineration · Structural funds

J. Kulczycka (✉)
AGH University of Science and Technology, Cracow, Poland
e-mail: kulczycka@meeri.pl

L. Lelek
Mineral and Energy Economy Research Institute of The Polish Academy of Sciences, Cracow, Poland

R. Salomone, G. Saija (eds.), *Pathways to Environmental Sustainability*,
DOI 10.1007/978-3-319-03826-1_19,

19.1 Introduction

One key element of EU policy is the continuous promotion of the implementation and development of environmentally friendly technologies (ETAP, SCP/SIP, Eco-Label, EMAS, IPPC, Energy-using Product Directive, GPP) which contribute to sustainable development. Environmental technologies play an important role in raising standards of living because they can decrease material inputs, reduce energy consumption and emissions, recover valuable by-products and minimize waste. They are considered as a strong driver to enhance environmental performance and to create new business opportunities.

In Poland investment projects contributing to sustainable development and which implement environmental technologies have been supported both from EU funds (Structural Funds) and also from national and regional funds. These funds are available for both private and public investors in the form of grants or loans, but their environmental and economic effects are taken into consideration in the evaluation process. Therefore specific sets of indicators were developed to measure the level of these effects. The environmental effect is based on comparison of the state before carrying out the investment and that after the investment, i.e. reduction in pollutants emitted into the atmosphere resulting from the construction of a new installation or the upgrading of an old one (Infrastructure and Environment 2007). The indicators are based on the Environmental Impact Assessment (EIA) methodology, therefore effects are calculated separately for each area of environmental protection (air, water, soil). This makes it difficult to compare ecological results (indicators) in different environmental fields. Moreover, this method does not take into account the entire life cycle of the investment.

One of tool which enables a comprehensive assessment to be made of the potential impact on the environment (taking into account all factors in the whole life cycle of an investment) is LCA (Life Cycle Assessment). This is the method recommended in many EU documents and which provides the opportunity to identify and priorities environmental threats, and thus search for technological solutions that maintain an optimal environmental quality. This method permits one to accurately identify all types of environmental damage and benefit in specific areas or technological phases and compare the ecological effects (Kulczycka 2009). Therefore the objective of this chapter is to analyze the indicators for evaluating environmental effects used by different environmental funds in Poland and to propose LCA (Life Cycle Assessment) as a tool for assessment of the environmental impact of investments supported by these funds.

19.2 Funds to Support Environmental Sustainability in Poland at the National and Regional Level

19.2.1 EU Funds at the National Level

Financial support tools for pro-environmental investment in Poland come *inter alia* from the EU Structural Funds. The main programmes helping to improve the quality of the environment in Poland and based on these funds include the Innovative Economy Operational Programme and the Operational Programme Infrastructure and Environment. Most of the investment projects implemented in 2007–2013 have been co-financed under these programmes.

The Innovative Economy Operational Programme was designed to manage the financial resources allocated to support innovation and eco-innovation solutions mostly in the production, marketing and organization stage. Direct and system support is provided for companies, business organizations and also for research institutes providing services for developing innovative businesses (Innovative Economy 2011). The purpose of the Operational Programme Infrastructure and Environment is to improve the investment attractiveness of the region, through activities related to the development of environmental infrastructure together with protecting and improving the quality of the environment and health, cultural identity and territorial cohesion (Infrastructure and Environment 2007). Under the programme the following support is provided (financial instruments):

- Funding of eligible expenses in the form of grants;
- Contribution of capital to venture capital funds;
- Funding in the form of a bonus technology which is part payment of the technological loan.

The beneficiaries of these programmes are academic institutions, research and industrial consortia, R&D units, Polish technology platforms, businesses (especially SMEs), public administration, higher education, institutions supporting the creation of innovative new businesses and groups of entrepreneurs, including: clusters, SMEs, large enterprises and technological networks.

The Regional Operational Programmes are the most important regional programmes financed by the EU in the period 2007–2013, and are implemented in 16 regions in Poland. These programmes relate to a wide range of social and economic activities. The main objectives include regional development, economic growth, and employment growth. The funds are intended to finance investments which are expected to contribute to improving the competitiveness and innovation of each region, and to improve intra-regional cohesion and the development of institutional capacity. The main projects funded under the programmes are:

- Investments in infrastructure,
- Investments to support innovation and the information society,
- Investments to improve environmental quality,
- Investments in the cultural sector.

19.2.2 National Funds

The National Fund of Environmental Protection and Water Management (NFEP&WM) was established in 1989 as a result of the regime transformation in Poland. Together with regional funds, it is the foundation of the Polish system of financing environmental protection. These funds provide financial support for projects which contribute to the enforcement of Polish obligations resulting from the Accession Treaty and other EU directives. The NFEP&WM manages its own finances as its main sources of income come from fees and fines for the use of the environment, service and concession fees, fees resulting from the application of the Energy Law, the Law on the Recycling of End-of-life Vehicles, income from the sales of Assigned Amount Units for greenhouse gas emissions and many other sources. The Fund's objective is to reduce the pollution of water, air and soil (National Fund 2012). The following forms of support are given by the funds:

- Loans (loans, credits granted by banks from the Fund, consortia e.g. joint financing Fund and banks, credit lines from Fund resources);
- Grants (investment grants, non-investment grants, subsidies to bank loans);
- Capital financing (acquisition of shares in newly created or existing companies in order to achieve an environmental effect).

Regional Funds similar to the NFEP&WM were also formed in each Polish province which support projects and investments in their specific region of the country. The purpose of the Fund is to provide financial support for projects designed to protect the environment and respect its values based on the constitutional principle of sustainable development while maintaining environmental safety and environmental programmes in the region concerned.

Other national projects are for example the action taken by the Ministry of Environment called Green Technology Accelerator GreenEvo initiative. Its goal is to promote Polish green technologies on foreign markets.

19.3 Methods for Calculating the Ecological Effect of Investment Projects Funded from EU and National Funds

The ecological effects of investment supported by the structural funds at regional and national level in Poland are in general calculated as a reduction in the quantity of pollution released into the environment before and after the project started. For most investments, the assessment of the environmental effects of activities related to the construction of a pro-environment installation and/or the modernization of an old one is carried out descriptively and separately for each area of environmental protection:

- Atmosphere (air);
- Water (sewage treatment plants, sewage systems);
- Road infrastructure;
- Land (waste management).

This makes it difficult to compare the results. Environmental impact assessment reports (EIAs) are mostly carried out for large investments due to the workload involved and their multi-stage nature. The actual methods used to calculate the ecological effects of projects related to the Infrastructure and Environment Programme (Infrastructure and Environment 2007) are presented below.

The ecological effects of projects are calculated separately for different types of protection. In case of air pollution the ecological effect is expressed as the difference between the initial and final (after the project) emission levels by type of pollutant such as SO_2, CO, NO_X, etc.):

$$E_r = \sum_{t=1}^{n} E_t * K_t$$

E_r Equivalent emission sources
t Number of different pollutants emitted from the source emission being evaluated
E_t The actual emission of pollutants with index t
K_t Toxicity factor of pollution with index t; this is thc ratio of the permissible annual average concentrations of SO_2 (eSO_2) to the limit of the annual average concentration of a pollutant (et)

$K_t = eSO_2/et$
K SO_2 = 30 mg/m^3: 30 mg/m^3 = 1.0 (toxicity factor)
K NO_x = 30 mg/m^3: 40 mg/m^3 = 0.75
K CO = 30 mg/m^3: not specified = not specified
K CO_2 = 30 mg/m^3: not specified = not specified
K Particulate = 30 mg/m^3: 40 mg/m^3 = 0.75
K Soot = 30 mg/m^3: 8 mg/m^3 = 3.75
K Benzo(a)pyrene = 30 mg/m^3: 0.001 mg/m^3 = 30,000 (Regulation of the Minister of the Environment 2002)

The ecological effect (EE) in the case of land management is based solely on the physical effect, and does not taken into account any environmental aspects. In calculating what the ecological effect is:

EE = P (Before-After)
P (Before-After): area of recovered degraded land [ha].

The ecological effect in the case of the effect on municipal wastes is estimated based on the physical effect (e.g., reduction in ton/years of municipal waste and a special factor) (Table 19.1):

$$E = \sum O_0 * w_0 - \sum O_1 * w_1$$

O_0 The amount of waste of a specific category subject to other disposal processes within the project [Mg/year]
O_1 After disposal processes [Mg/year]
w_0 Harmfulness ratio of waste before disposal, classified in accordance with the Waste Catalogue
w_1 After disposal

Table 19.1 Harmfulness ratio of waste

Type of Waste	Harmfulness ratio of waste before disposal	Harmfulness ratio of waste after disposal
Inert waste	0.11	0.10
Waste other than inert and hazardous	0.50	0.45
Hazardous waste	5.00	4.50

In the national project (Operational Programme Infrastructure and Environment), the environmental effect for this type of impact is calculated using the cost effectiveness indicator for environmental effect (WK):

$$WK = \frac{\sum_{i=1}^{i=n} c_i * O_i * Z_i}{ZKK + RKE}$$

O_i Unit base rate fee for use of the environment for "i"
Z Amount of the reduced factor "i" in the first year after the investment
i Index of impact factor
n Number of impact factors included in the calculation
ZKK Investment expenditure annualised
RKE The annual operating costs of installation c_i (correction coefficient, the following values are adopted, $c_i = 10$ for hazardous waste, $c_i = 1$ for other than hazardous waste and for water and energy savings)

The methods presented are relatively simple and can be treated as tools for promoting sustainable development. However they are not suitable for solutions for the eco-innovation and environmental technology fund projects which should take into account all impacts during the product life cycle.

19.4 Methods for Calculating the Ecological Effect of Investment Projects Funded from EU and National Funds

The IWP-PL model was developed in order to promote and implement life cycle thinking in Poland. It is a tool designed to provide support at the design stage to the decision-making process on investment related to municipal waste management systems. Developed on the basis of a computer program, IWM-PL helps the user to evaluate different scenarios and all waste management systems using LCA (*Life Cycle Assessment*). The program is a combination of the IWM-2 model and the Eco-indicator 99 method. IWM-2 was the basis for the analysis of ecological effects in the form of emissions to air and water. Eco-indicator 99 was used to model environmental impact assessment in the so-called impact and damage categories.

The overall assessment of the municipal waste management system includes an: environmental analysis (in accordance ISO 14040:2009) and an economic analysis (based on dynamic methods using net present value).

The final results depend on the date introduced by the users who should have basic knowledge about the project as there are 28 steps which need to be completed to obtain the final economic and ecological results. However the results are only presented in 6 impact categories which include carcinogens, respiratory organic and inorganic, climate change, ecotoxicity, and acidification/eutrophication. The ecological results are shown as so-called "ecopoints" (Pt), defined as the ratio of the total annual burden on the environment (emissions, resource consumption, land use) in Europe to the population, then multiplied by 1,000. In this way the rate of 1,000 Pt corresponds to an annual environmental load caused by the average citizen in Europe throughout their life cycle. The higher the number, the higher the potential impact of the process or product analyzed on the environment (Eco-indicator 99 2009). The economic results, by contrast, show the efficiency of financial investments and these are presented in accordance with the methodology presented for European investment projects (Guide to Cost-Benefit Analysis 2008).

IWM-PL allows to model many municipal waste management scenarios using different technologies, such as sorting, composting, biogasification, the mechanical-biological treatment process, thermal treatment or landfill.

19.4.1 Case Study Using IWM-PL for the Environmental Assessment of Municipal Waste Incinerators in Warsaw

Municipal Waste Incineration in Warsaw commenced in 2000. In addition to municipal waste disposal by thermal treatment, the plant carries out processes concerned with waste segregation, the recovery of recyclable materials, composting a separate organic fraction, neutralization of products generated during the combustion process, ash and slag in the form of environmentally neutral granulate, and energy production. Some of the energy produced in the combustion process media is used for the plant's own company, while the surplus is sold to the grid.

The objective is to calculate the annual environmental effect for waste managed at Warsaw's incineration plant and to compare it to landfilling using the IWM-PL program. The scenario represents the current state of the plant (with energy recovery during the combustion process at the level of 30 %). The functional unit adopted for an annual intake of municipal waste, and the system boundaries include the plant life cycle processes (not including the life cycle of the infrastructure, land use, and transportation), i.e. sorting, incineration (landfill in Scenario II), composting, recycling, disposal of fly ash and slag. Data for the analysis came from materials provided by the owner of the Warsaw Plant and IWM-PL databases. The analysis was performed taking into account the balance of materials, fuel and energy, the number of recyclable materials, the process of composting the biodegradable fraction, the combustion process and the management of ash and bottom slag (Table 19.2).

Table 19.2 Inventory data used in the environmental analysis of Municipal Waste Incineration in Warsaw. (Source: Data obtained from the owner of the Warsaw plant)

No.	Inputs	Amount	Unit
1	The amount of waste taken to the plant	62 748	Mg
2	Amount of recyclable materials recovered through segregation	1 225	Mg
3	The composition of recyclable materials:		
	Glass	196	Mg
	Ferrous metals	735	Mg
	Non-ferrous metals	294	Mg
4	Amount of metals recovered from ashes	293	Mg
5	Amount of biodegradable waste for the composting process	18 712	Mg
6	Amount of mixed municipal waste sent to landfill or the incineration process	42 518	Mg
7	Energy recovery in the incineration process	30	%

Assessment of the environmental risks and benefits associated with waste management in Municipal Waste Incineration in Warsaw were presented in 6 impact categories:

- Carcinogens,
- Impact of inorganics compounds on human respiratory system,
- Impact of organics compounds on human respiratory system,
- Climate change,
- Acidification/eutrophication,
- Eco-toxicity (Fig. 19.1),

and two damage categories:

- Human health (sum of the carcinogens, resp. inorganics, resp. organics, climate change),
- Ecosystem quality (sum of the acidification/eutrophication and eco-toxicity) (Fig. 19.2).

Energy recovery from the waste combustion process (at level 30 %) makes the potential impact on the environment have negative values-to the benefit of the environment (Fig. 19.1). This is because to "replace" a significant amount of energy (electricity produced from conventional sources) in the grid by the energy recovered from the waste combustion process, results in a reduction of emissions of SO_x, NO_x and particulates. These emissions mainly have an impact on the human respiratory system, therefore, the impact category "resp. inorganics" and damage category of "human health" is dominant (Fig. 19.2). The highest environmental load is caused by emissions of CO_2, which reflects the "climate change" impact category (Fig. 19.1). However, this impact is completely compensates by the environmental benefits (Fig. 19.2).

The results for landfilling (the same amount of waste) amounted to about 283 000 Pt, therefore the environmental effect can be calculated as the difference between these two results (−664 580.21 and 283 000 Pt).

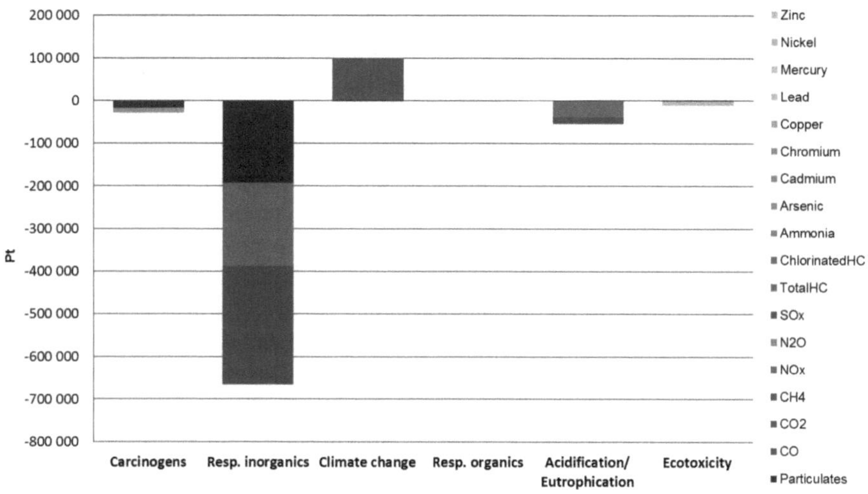

Fig. 19.1 The results of the environmental analysis (air emissions) for Municipal Waste Incineration in Warsaw in six impact categories (Scenario I). (Source: Own calculations using the IWM-PL)

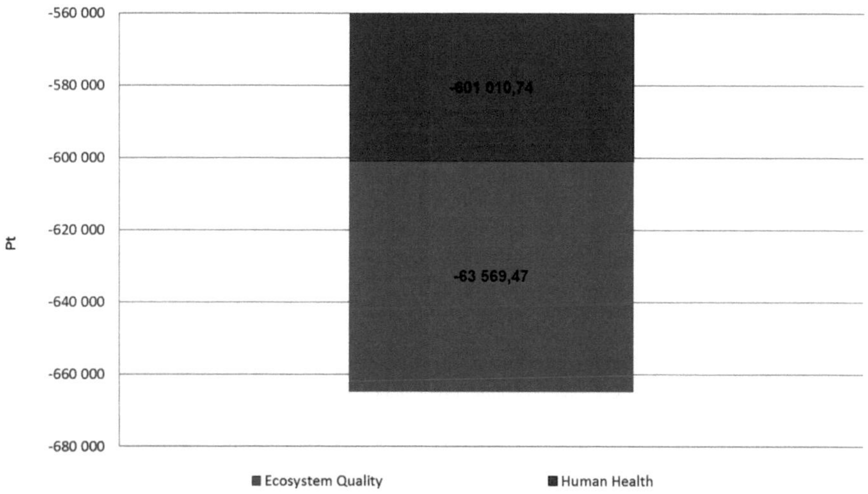

Fig. 19.2 The results of the environmental analysis for Municipal Waste Incineration in Warsaw in two damage categories (Scenario I). (Source: Own calculations using the IWM-PL)

19.5 Summary

Programs that use the LCA method for modeling waste management systems have already been developed in some countries such as the UK and in the main they are widely used to identify the potential risks and environmental burdens, and the benefits from different investment options. The IWM-PL software is the first available in the Polish language that allows the use of LCA in this area. The IWM-PL program is based on the IWM-2 model, and the method Eco-indicator 99 allows one to present the results in the impact and damage categories but takes the influence of the composition of waste into account. The software is complemented by aspects of the economic efficiency of investment, allowing one to develop a model of the system analyzed in an easy and effective manner, and then to determine its potential impact on the environment and economic viability. Simulations presented in the example shown "environmental benefits" resulting from incineration of waste compare to their landfilling. This is only a presentation of one of the scenarios, but the program allows the construction of various alternative solutions and so one can identify the effects of both environmental and economic factors. Therefore, the software can be used by local authorities and potential investors to make investment decisions because the results focus on the key environmental and economical factors taking into account the life cycle of the investment (Kulczycka 2009).

References

Act of 3 October (2008) on the provision of information about the environment and its protection, public participation in environmental protection and environmental impact. Journal of Laws No. 199, item. 1,227

Com (2011) Communication from the Commission to The European Parliament, the Council, the European Economic and Social Committee and the Committee of The Regions, Roadmap to a Resource Efficient Europe. 571, Brussels

Eco-indicator 99 (2009) Manual for Designers. PRé Consultants. http://www.pre-sustainability.com/download/manuals/EI99_Manual.pdf. Accessed 16 Nov 2012

Guide to Cost-Benefit Analysis (2008) Guide to Cost-Benefit Analysis of investment projects: Structural Funds, Cohesion Fund and Instrument for Pre-Accession. European Commission, Directorate General Regional Policy. http://ec.europa.eu/regional_policy/sources/docgener/guides/cost/guide2008_en.pdf. Accessed 16 Nov 2012

Infrastructure and Environment (2007) Infrastructure and Environment Operational Programme, National Strategic Reference Framework 2007–2013 (2007), Ministry of Regional Development. http://www.mrr.gov.pl/fundusze/wytyczne_mrr/konsultacje/horyzontalne/Documents/20120425_wytyczne_sektor_odpadowy_aktualizacja_po_kon.pdf. Accessed 16 Nov 2012

Innovative Economy (2011) Innovative Economy Operational Programme, the National Strategic Reference Framework 2007-2013 (2011). http://www.poig.gov.pl/Dokumenty/Lists/Dokumenty%20programowe/Attachments/116/Program_Innowacyjna_Gospodarka_zatwierdzony_przez_KE_22122011.pdf. Accessed 16 Nov 2012

ISO (14040:2009) Environmental management - Life cycle assessment - Principles and framework. http://www.iso.org/iso/catalogue_detail?csnumber=37456. Accessed 16 Nov 2012

Kraszewski A, Pietrzyk-Sokulska E (2011) Assessment of waste management system. IGSMiE PAN, Krakow

Kulczycka J (2009) Life cycle thinking in Polish official documents and researches The determination of discount rate for green public procurement. Int J Life Cycle Assess 14:375–378

Kulczycka J, Pietrzyk-Sokulska E, Góralczyk M, Koneczna R, Spielmann M, Merl A (2008) The development of LCA methodology for the evaluation of infrastructure projects. IGSMiE, PAN, Krakow

National Fund (2012) National Fund for Environmental Protection and Water Management. http://www.nfosigw.gov.pl/. Accessed 16 Nov 2012

Piszczek J, Osicki A, Kukla P (2010) Air Protection programs, programs to improve air quality, low-emission reduction programs, methods of calculating the baseline and environmental effects. Foundation for Energy Efficiency, Katowice

Regulation of the Minister of the Environment (2002, 6 June) on permissible concentrations of air pollutants, The National Strategic Reference Framework 2007-2013, Guidelines on selected issues related to the preparation of investment projects, including revenue-generating projects (2009), Ministry of Regional Development. http://www.mrr.gov.pl/fundusze/wytyczne_mrr/obowiazujace/horyzontalne/Documents/wytyczne_w_zakresie_wybranych_zagadnien_zwiazanych_z_przygotowaniem_projektow_inwestycyjnych.pdf. Accessed 16 Nov 2012

Chapter 20
The Project for the Implementation of the Industrial Symbiosis Platform in Sicily: The Progress After the First Year of Operation

Laura Cutaia, Roberto Morabito, Grazia Barberio, Erika Mancuso, Claudia Brunori, Pasquale Spezzano, Antonio Mione, Camillo Mungiguerra, Ornella Li Rosi and Francesco Cappello

Abstract In May 2011, the Environmental Technologies Technical Unit of ENEA launched a project to build a platform of Industrial Symbiosis to be implemented in Sicilia Region. It is a 3 years project, in a framework of a larger asset to support productive development in southern Italy. This chapter summarizes the activities carried out during the 1st year of operation by the working group that participates in the initiative. The platform, whose main objective is to help in launching industrial symbiosis through a geo-referred information system support, acts as a tool in the service of business and territory and also offers a range of tools that may be of interest and use especially for SMEs. In addition to the technical aspects of the implementation of the information, tools, and website, the activities involved a first phase of consultation in Sicily with stakeholders, with particular reference to Confindustria Sicilia—with which ENEA has signed a specific Cooperation Agreement—and Sicilia Region, Regional Agency for Waste.

Keywords Industrial symbiosis · Network · Sicily · GIS · Input–output

L. Cutaia (✉) · R. Morabito · G. Barberio · E. Mancuso · C. Brunori · P. Spezzano
Environmental Technologies Technical Unit, ENEA, Roma, Italy
e-mail: laura.cutaia@enea.it

A. Mione
Environmental Technologies Technical Unit, ENEA, Pisa, Italy

C. Mungiguerra
Technical Unit for Information Systems and ICT Development, ENEA, Roma, Italy

O. Li Rosi
Technical Unit for Advanced Technologies for Energy and Industry, ENEA, Ispra, Varese, Italy

F. Cappello
Energy Counseling Centre Sicilia, ENEA, Palermo, Italy

R. Salomone, G. Saija (eds.), *Pathways to Environmental Sustainability,*
DOI 10.1007/978-3-319-03826-1_20,

20.1 Introduction

Industrial ecology is based, since 1950s, on the evidence that industrial system can be described as distribution of materials, energy and information flows and there is exchange between the industrial system and the natural system (Garner and Keoleian 1995; Erkman 1997). Industrial symbiosis (IS) is part of industrial ecology that operates from the global level to the level of individual firms and different definitions of IS can be find in literature to identify this kind of cross-firm relationship. The first definition of IS comes from Renner (Renner 1947) as "relationships between industries, sometimes simple, but often quite complex, which enter into and complicate the analysis. Chief among these is the phenomenon of industrial symbiosis. By this is meant the consorting together of two or more of dissimilar industries."

One of the most recent definitions (Lombardi and Laybourn 2012), industrial symbiosis (IS): "*engages diverse organisations in a network to foster ecoinnovation and long-term culture change. Creating and sharing knowledge through the network yields mutually profitable transactions for novel sourcing of required inputs, value-added destinations for non-product outputs, and improved business and technical processes.*"

This definition was anticipated, always by the same authors, earlier in 2010, in a slightly different form: "*IS engages traditionally separate industries and other organisations in a network to foster innovative strategies for more sustainable resource use (including materials, energy, water, assets, expertise, logistics etc.) Through the network, business opportunities are identified leading to mutually advantageous transactions for innovative sourcing of required inputs, and value-added destinations for non-product outputs.*"

The proposed definitions differ from the one proposed by Chertow (2000): "*The part of industrial ecology known as industrial symbiosis engages traditionally separate industries in a collective approach to competitive advantage involving physical exchange of materials, energy, water and by-products. The keys to industrial symbiosis are collaboration and the synergistic possibilities offered by geographic proximity*"; it also differs from the next definition by Jacobsen in article describing the specific case of Kalundborg as particular case that match theoretical principles of IS and industrial ecology into practice and quantifying the environmental and economic direct and indirect benefits: "*IS can be categorized as a concept of collective resource optimization based on by-product exchanges and utility sharing among different collocated facilities.*"

One of the main points of divergence between the reported definitions is in the "*geographic proximity*" of Chertow which effectively becomes "*economic proximity*" to Lombardi and Laybourn. Sharing this approach, the industrial symbiosis can occur where there is evidence synergies in an "appropriate" economic and geographic proximity, that is, where it is possible to identify opportunities not only from a technical point of view, but also logistical and operational, as well as from the economic point of view compared to the previous "not symbiotic" solutions.

At international level there are some significant initiative, among those one of the most relevant is the NISP (National Industrial Symbiosis Programme of UK—www.nispnetwork.com). It works according with the IS network approach, using the Synergie™ software developed within the program started in 2005. There are some experiences based on the same network approach, for instance the e-SYMBIOSIS LIFE+Project (www.esymbiosis.gr) started in 2010 and lasting 36 months.

At Italian level, one of the most significant public action toward industrial symbiosis at regulatory level can be detected in the so-called Bassanini Decree which in 1998 establish the ecologically equipped productive areas (Art. 26 D.Lgs. 112/98), even if up to know the decree has not been adopted by all the Italian regions. But this decree is more addressed to the realization of common services and management in industrial areas, rather than to specific IS connections. In the recent past, Italian regulation moved toward concepts closer to IS adopting the law 133/2008 concerning the "enterprises networks" as "open aggregations of individual production centers joined for the development of industrial policies"; in 2010 this regulation has been updated through the Art. 42 law 122/2010. The approach of "enterprises networks" seems to be more suitable to be direct toward the set-up of connections, even symbiotic, between companies.

This chapter shows the activities carried out for the implementation of an Italian Industrial Symbiosis Platform project launched in May 2011 by the Environmental Technologies Technical Unit (UTTAMB) of ENEA (Roma, Italy). The implementation is focused on the Sicilia Region (Cutaia et al. 2011), as part of the wider ENEA project "Eco-innovation Sicilia" (Brunori et al. 2012) which foresees, among other things, the realization of two main activities: "Sustainability of production systems in the Sicilia Region: a pilot project in the field of Plastic and Electronic Equipment" (WP1) and the "Sustainable Tourism: a pilot intervention in the Archipelago of the Egadi Islands" (WP2). WP1 includes three sub-activities: Technologies for the recovery of raw materials and for the integrated management of electronic waste (WEEE); technologies for the exploitation and recycling of plastics coming from WEEE; Development and implementation of a regional industrial symbiosis platform implemented, in particular, to the field of WEEE and plastics. The IS Platform is designed and developed to enable synergies of any type of flow, even if, for integration with other actions of the project, will initially be implemented for WEEE and plastics. Furthermore is based on a Geographic Information System (GIS) allowing to compare and share many other information layers and tools, as the Platform is intended to be a tool-set available for Italian companies, SMEs in particular.

20.2 Industrial Symbiosis and the Start-Up of the Project

Starting from the technical–scientific reference (Cutaia and Morabito 2012), the industrial symbiosis platform has taken start from one of the most recent definitions of industrial symbiosis (Lombardi and Laybourn 2012) above mentioned. So, it is possible to identify a number of key elements that form the structure of the ENEA Platform:

- The network: the possibility to bring together different stakeholders, i.e., the actions needed to enable the networking of stakeholders
- Information: knowledge of the opportunities available in an area both in terms of its industrial features and services, both in terms of availability of specific resources that can be shared to enable synergies between different stakeholders
  - General DBs, also geo-referred, describing the local framework
  - Specific DBs on the availability of resources as output and the request for raw materials as input: such information requires the collaboration of individuals and should therefore be co-operative
- Skills: including technical and scientific, which allow to identify possible synergies (the operator of the platform)
- The participation of users: the interface with potential users through a web portal and dissemination and promotion activities

The Industrial symbiosis platform is therefore proposed as a tool to create synergies according to the network approach: identify opportunities for synergies between different actors in a way that waste/by-products of one can become an input for the other, without the need to activate exchanges of resources exclusive, continuous and long-lasting, like e.g., Kalundborg (Jacobsen 2006).

In addition to the specific function of activating industrial symbiosis pathways, the platform, as part of a project to support productive activities, will offer a tool set for companies, for SMEs in particular, and in general, for any kind of user. Among these, as set forth below, will be provided information tools for consultation of environmental regulations and best practices, geo-referred DBs, which can be, at certain conditions, viewed and queried by users. In addition to these, the platform reports also two on-line tools to perform, in a simplified way LCA [Life Cycle Assessment, eVerdEE: Simplified Life Cycle Assessment (LCA) Tool for SMEs] and Eco-design (TESPI: Tool for Environmental Sound Product Innovation) useful in particular for SMEs. These tools have been developed and are managed by the Technical Unit for Models, Methods and Technologies for Environmental Assessments—LCA and Ecodesign lab. of ENEA located in Bologna (Buttol et al. 2011).

20.3 Progress of the Project

Up to now (end 2012), the project has carried out the following major activities:

- The start of stakeholders involvement in Sicilia for network activation
- The creation and registration of the logo "Symbiosis"
- The creation of the website for the promotion and dissemination of activities, to allow stakeholders to join
- The design and implementation of the IT architecture of the platform, tools, and data collection

SYMBIOSIS
Piattaforma di simbiosi industriale

Fig. 20.1 Logo of the ENEA Industrial Symbiosis Platform

In addition to activities mentioned above and illustrated in the following, the working group is carrying out other aspects of the design including the geo-referencing and characterization of valorization and treatment plant, particularly referring to those for treatment of WEEE and plastics, as well as a DB "string origin-destination" "by-product/waste, productive use" and vice-versa, developed according to the logic one-to-many. Examples of cards of productive reuse of industrial by-products are already available, as an example, on the website. There are also still in the development stage the input-output data collection sheets for the specific information of associated users which have to feed their own input-output sheet on the specific cooperative information layers (Fig. 20.3).

20.3.1 The Start of the Stakeholders' Involvement

Between 2011 and 2012, contacts were established with the Sicilia Region and Confindustria Sicilia. For the Sicilia Region in particular with the Regional Waste Agency, with regard to the census of first and second level treatment and valorization plants for waste and by-products.

At the same time there were various consultation meetings with Confindustria Sicilia (Confindustria is the main organization representing Italian manufacturing and services companies) which led, June 4, 2012, the signing of a Framework Agreement for eco-innovation in the productive systems (Adnkronos 2012), including industrial symbiosis. Next steps for promoting the Platform and involving users in Sicilia will foresee a set of specific local meetings.

20.3.2 The Creation and Registration of the Logo

In order to identify the IS Platform, was developed the logo "Symbiosis" (Fig. 20.1) on the basis of a specific graphic design. This graphic is used for the project's website and for any informative and promotional material. The logo can be used, as in the case of project Ecoinnovazione-Sicilia, with the insertion, at the bottom right, of the name of the region in which the activity is implemented (in this case "Sicilia").

The logo was subsequently recorded in its various forms, with Community Trade Mark ENEA ref. 744—CTM FIGURATIVE (Symbiosis) No. 10517258."

The Registration of the logo with ENEA as owner, allows its use even in the case of subsequent implementations and extensions of the platform to other regions.

20.3.3 The Web Site Implementation

In September 2011 ENEA has registered the domain *www.industrialsymbiosis.it* and other equivalent, as the project website. As well as for the logo, the "owner" of the domain is ENEA and therefore can use it both for Ecoinnovation-Sicilia project and for any other application of IS Platform to other region and projects carried out by ENEA. In November 2012, during the Ecomondo 2012 trade fair (Cutaia 2012), ENEA has organized and sponsored a conference on the role of industrial symbiosis for the green economy and the project website has been put online and is currently available, even if only partially complete because the project is still on-going.

The project website is built using the Plone technology (Marchetti 2005), an open-source Content Management System (CMS): "a CMS is a tool that allows the centralized and decentralized editors to create, edit, manage and finally publish on the web a large number of content while these are controlled by a centralized set of rules, processes and controls flow (workflow) that ensure an appearance web validated and coherent."

The Plone technology is well suited for application in projects of this type for a variety of reasons, including the ease of content management, the ability to search and index the information, the possibility of applying rules about who can publish what and when, the ability to integrate with other systems. This last feature makes it suitable, in particular, to be interfaced with other systems that, together, make up the informatics architecture of the platform.

Moreover, the choice to use open-source software, also performed for the architecture of the platform is in line with the guidelines of the Ministry of Innovation and Technology, which, already in 2003, issued a directive for the "Development and use computer programs by public administrations" (Italian Official Gazette of February 7, 2004), that invite public bodies to consider the possibility to use "open-source" computer programs.

20.3.4 The Informatics Architecture of the Platform

The Fig. 20.2 shows the block diagram of the informatics architecture of the platform. This is the modular composition of open-source software that constitute respectively the Desktop Manager software through a GIS (Geographic Information System, software Quantum GIS—www.qgis.org), the DB server, i.e., the server for the publication of cartographic files (shape and raster) and for geo-location, as well as the application server for all management services for the management of DBs and for their query and for connection with users via web.

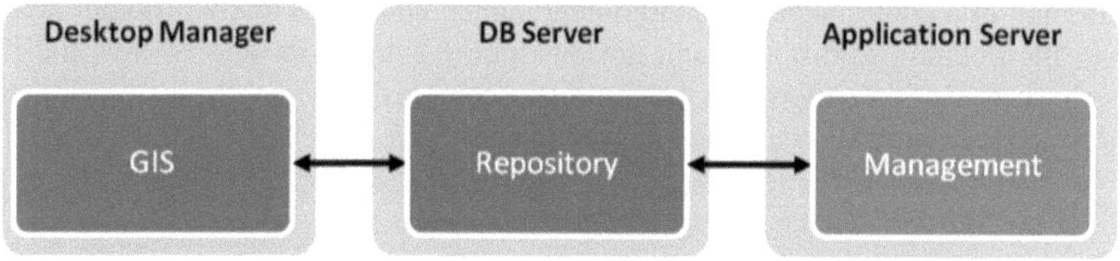

Fig. 20.2 Block diagram of the informatics architecture of the platform

Without going into details of all the specific software used, it is interesting to note once again that the architecture of the whole platform is based on open-source software, including map-based reference (www.openlayers.org) making in this way the platform itself able to adapt to changes in both technological and spatial aspects and details.

It is also on-going the DB and reports development for put in communication the various layers of information in order to identify potential synergies between two or more points in the network (users) through the possible input-output relationships. In addition, the activity of definition of the repository of the DB "string origin-destination," "by-product/waste, productive use" and vice-versa, developed according to the logic one-to-many, that allow the detection of the possible selections of synergies between two or more point of the network.

20.3.5 Users

The operation of the platform requires collaboration between users (companies, local authorities) and the experts who manage and implement its website (Portal). In fact, the information system, which is the knowledge base, is a dynamic tool (that is updated periodically by the operator) and cooperative (i.e., supplemented with specific information provided by users). Through the portal stakeholders can become members taking advantage of different degrees of membership:

- Registered users, who access to a set of basic tools
- Associated users, who access to all the tools and simultaneously cooperate in the enrichment of the geo-referenced DB, providing their own specific information about the resources used in their industrial activity, as well as waste and/or by-products generated. Are users who can activate specific queries to identify possible industrial symbiosis connections.

In addition to the above mentioned categories of users, there are also cooperative structures, i.e., macro-users, providing one or more specific informative layers (e.g., associated list within a local industry association, ecc): these layers become part of the DB of the platform but its owner is the cooperative structure. This mechanism is a mutual enrichment for both the platform (which can have specific and updated informative layers), both for the cooperative structure (that can take advantage of a

GIS that, in a coordinated and congruent manner, composes many different layers that often belong to different and not dialoguing actors). In other words, the IS-Platform is also an opportunity to compile a complex GIS sharing many different informative layers, allowing the integration and the cross-relation between them.

20.3.6 The Development of the First Tools

20.3.6.1 Regulatory and Administrative Framework and Best Practices

A key issue for users of the platform of industrial symbiosis is the *regulatory framework* that establishes the legal boundaries within which to move. To this end, we proceeded to the definition of a framework that shows the possible law topics useful for exchanging resources between users, such as the "reuse," the "recovery," the "end of waste," the "by-product," and "secondary raw materials" concepts. These concepts offer to the companies, but in general to all the stakeholders, the opportunity to replace the traditional management model of the "cycle of waste" with a new "value chain" of the materials. The intention of the European legislator is in fact in favor of this exchange of resources, as foresee by waste hierarchy (Dir. 2008/98/EC) with the concepts of "waste prevention" and "preparation for re-use." It is translated at national level in the new "waste" definition and in the new concept of "end of waste" (Leg. Decr. 205/2010), and insists with the various changes to the old Leg. Decr. 152/2006 on the classification of "by-product." The proposed regulatory framework, therefore, is the result of a specific selection of regulations at European and national level that introduce and manage these concepts. As a tool for users will be soon available a selected DBs for consultation, providing main information about laws and regulations related with the IS concerns. The information provided in this section cannot substitute, in any case, authorization acts.

Further, the Platform intends offer users, mainly SMEs, the cognitive tools to access the *best available techniques and practices* for the improvement of their industrial processes, mitigation of environmental impacts, optimizing the use of resources. The European Union since 1996, through the IPPC Directive (96/61/EC), provides the best available techniques (BAT, Best Available Techniques) at Community level through the development of reference documents for different industrial sectors (BAT Reference documents—BRef). ENEA, with the intention of making the BRef easier to consult and use is building a user-friendly BRef DB identified by Italian keywords and abstracts and adding, where available, the Italian summary. The Platform also shares national guidelines, as transposition of BRef by Italian authorities.

20.3.7 The Hierarchy of Operation and the Databases

The Fig. 20.3 shows the hierarchy of the Platform that from the point of view of the information is based on geographic information layers that represent the technical

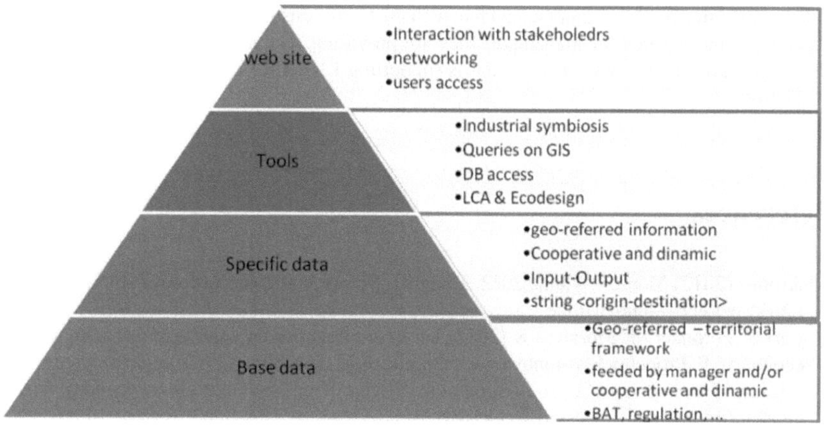

Fig. 20.3 Hierarchy of the ENEA IS Platform

and logistical information of an area (base data) and the specific streams of input and output of users. Up to now it has been possible to begin to acquire a first set of information layers concerning, for example, the registry of interested parties, the waste disposal plants, platforms, raw and secondary material treatment plants. However, this information have been provided or found with incompatible formats for their up-load on GIS ENEA. Among the auspices of the project is also to be able to activate "dynamic links" between existing DBs (e.g., managed by local authorities). As a "second best" can be activated "static links" to existing DBs, possibly agreeing regular updates with the data owner. As last option is the manual effort to fill in the GIS. According to the mechanism of the "cooperative structures" discussed earlier about the different types of users, the sharing of information layers on the Platform can be of mutual benefit to the Platform itself and to the "owner" of the information layer.

20.4 Conclusions and Next Steps

Activities carried out so far have registered the interest of the first institutional stakeholders and industry associations that have been involved. In the near future, however, it is foreseen a set of meeting with local stakeholders who are, actually, the users of the Platform. At the same time it is needed to finalize the registration user's protocol and data collections both for associated and cooperative users. The development of "origin, destination" strings as well as characterization of valorization and treatment plants are on-going activities, addressed the first at connect output from one user to the input of one other (not on GIS) and the second at the whole description of treatment plants in the selected area (on GIS).

Acknowledgments The authors would like to thank Confindustria Sicilia and the Sicilia Region—Waste Regional Agency for the support they are providing. The authors also wish to thank Federico Guizzardi of Abstract S.R.L. which is supporting ENEA for the informatics architecture of the Platform.

References

Adnkronos (2012) Monday, 4 June 2012, 18.29.07, Sicilia: siglato accordo CONFINDUSTRIA—ENEA per eco—innovazione

Brunori C, Peronaci M, Morabito R (2012) Un caso pilota per un approccio integrato alla green economy. Il Progetto Eco-innovazione Sicilia, EAI Speciale I-2012 Verso la green economy. http://www.enea.it/it/produzione-scientifica/pdf-eai/speciale-green-economy/specialegreeneconomy.pdf. Accessed 26 Oct 2012

Buttol P, Buonamici R, Naldesi L, Rinaldi C, Zamagni A, Masoni P (2011) Integrating services and tools in an ICT platform to support eco-innovation in SMEs. Clean Technol Environ Policy. doi:10.1007/s10098-011-0388-7 (Online first, 26 May)

Chertow MR (2000) Industrial symbiosis: literature and taxonomy. Ann Rev Energy Environ 25:313–337

Cutaia L (2012) Il progetto per lo sviluppo della Piattaforma Regionale di Simbiosi Industriale. La simbiosi industriale quale strumento per la Green Economy Conference (8.11.2012), Ecomondo 2012, Maggioli Editore, Santarcangelo di Romagna, Italy

Cutaia L, Barberio G, Li Rosi O, Mancuso E (2011) Verso una piattaforma di simbiosi industriale: l'attività in corso nell'ambito del progetto Eco-innovazione Sicilia (Developing industrial symbiosis platform: activities in the Sicilia Ecoinnovation project). Ecomondo Proceedings, Rimini, November 2011, Maggioli Editore, Santarcangelo di Romagna, Italy

Cutaia L, Morabito R (2012) Ruolo della Simbiosi industriale per la green economy. Uno strumento innovativo per la chiusura dei cicli delle risorse. EAI Speciale I-2012 Verso la green economy. http://www.enea.it/it/produzione-scientifica/pdf-eai/speciale-green-economy/specialegreeneconomy.pdf. Accessed 26 Oct 2012

Erkman S (1997) Industrial ecology: an historical view. J Clean Prod 5(1–2):1–10

Garner A, Keoleian GA (1995) Industrial ecology: an introduction. Pollution prevention and industrial ecology. National Pollution Prevention Center for Higher Education, University of Michigan

Jacobsen NB (2006) Industrial symbiosis in Kalundborg, Denmark. J Ind Ecol 10(1–2):239–255

Lombardi R, Laybourn P (2012) Redefining industrial symbiosis. Crossing academic–practitioner boundaries. J Ind Ecol 16(1):28–36

Marchetti S (2005) Plone: il CMS ideale per PA e PMI, Ferrara, Linux Day 2005

Renner GT (1947) Geography of industrial localization. Econ Geogr 23(3):167–189

Chapter 21
Potential Developments of Industrial Symbiosis in the Taranto Productive District

Bruno Notarnicola, Giuseppe Tassielli and Pietro Alexander Renzulli

Abstract Industrial Symbiosis represents one of the approaches of the Industrial Ecology paradigm implemented as a means of making industrial production more sustainable. Currently in Italy there are only a few initiatives entailing Industrial Symbiosis and they have not yet developed fully. This chapter represents a study of the industrial sector of the Taranto province located in south east Italy, whose aim is that of identifying current and some new potential Industrial Symbiosis interactions among the firms of the sector. The economic, environmental, waste and energy analysis of the provincial industrial sector, has depicted the current state of Industrial Symbiosis and has also identified who the key players of a future Industrial Symbiosis scenario in the province could be. The waste reutilization and energy management schemes identified and described in this chapter, regarding the yearly generation of 3.25 Mt of solid waste and by products and over 1,000 ktoe of waste heat, could be used as a starting point for a real implementation of Industrial Symbiosis interactions that will effectively be able to make the Taranto province industrial system more competitive and environmentally sustainable.

Keywords Industrial ecology · Industrial symbiosis · Eco-industrial parks · Taranto productive district · Waste and energy analysis

21.1 Introduction

Current global levels and patterns of consumption of modern day societies are unquestionably unsustainable in environmental-social terms and often also in economic terms. This calls for a review of consumption patterns and policies together with production approaches. The Industrial Ecology (IE) paradigm, over the last decades, has involved the study and application of new approaches regarding industrial production with a specific objective of attaining sustainable development. One such IE approach is Industrial Symbiosis (IS), that entails minimizing toxic

B. Notarnicola (✉) · G. Tassielli · P. A. Renzulli
Ionian Department of Law, Economics and Environment, University of Bari "Aldo Moro",
Via Lago Maggiore angolo via Ancona, 74100, Taranto, Italy
e-mail: bruno.notarnicola@uniba.it

R. Salomone, G. Saija (eds.), *Pathways to Environmental Sustainability*,
DOI 10.1007/978-3-319-03826-1_21, © Springer International Publishing Switzerland 2014

materials and emissions of waste and pollutants over the life cycle of products and increasing the efficiency of energy and matter use via the creation of strategic and synergic industrial cooperation both at a local and a wide area level.

In Italy there are currently several IS projects, such as the one in the Pistoia-Prato-Lucca districts (Dinelli et al. 2003), the one in Sangro-Aventino industrial area (Romano et al. 2010) and the one in the Sicily region (Cutaia et al. 2011). All are still under development and have not yet progressed to stable symbiotic interactions. Current Italian state legislation for development of productive areas, called Aree Produttive Ecologicamente Attrezzate (APEA), tends to include more IS concepts in order to make such districts more sustainable. However the application of such concepts is transferred to regional governments and only a few of these have developed specific legislation. The Puglia Region has included indications for the APEA in its regional environmental protection plan (PPTR) which is not yet in force. The Distretto Produttivo per l'Ambente e Riutilizzo (DIPAR), located in Taranto, is an organization whose aim is that of developing a network, among local firms and local government organizations operating in the area, to be used as a means of exchanging information and promoting best practices employed to reuse waste in a sustainable way, but it is currently still under development.

The objective of this chapter is that of performing a study, with particular reference to IS approaches, of the industrial sector of the Taranto province located in south east Italy in the Apulia region. Specifically this chapter aims to identify hot-spots and critical issues of the provincial industrial district and also aims to identify current and some new possible IS interactions among the firms of the district, in order to develop background information that could be used in the future as a first step toward a practical implementation of IS principles within the province. In order to achieve the above mentioned objectives the work described in this chapter was organized as follows. The first stage involved a study of the literature related to IS with the aim of developing an idea of what is intended by IS, how it can be developed and applied to industrial districts. Following this first stage, an economic and environmental study of productive district was performed in order to identify who the potential actors could be in a IS scenario applied to the district. Next waste and energy analysis were performed. Questionnaires were then developed as a means of gathering further information to complete the picture regarding energy and waste. Based on the collected information, some possible synergic-symbiotic interactions to optimize the exchange of matter and energy, for the industrial district, were developed. Finally some preliminary ideas of further development of IS in the district were formulated.

What follows is a detailed description of the adopted approaches and methodologies together with the results obtained during the study.

21.2 Literature Review

Ever since the first recognized implementation of IS approaches which occurred in Kalundborg, a Danish industrial area, IS has been extensively studied and a large amount of literature has been produced (Boons et al. 2011). Generally speaking

IS engages traditionally separate industries in a collective approach to competitive advantage involving physical exchange of materials, energy, water, and/or by-products (Chertow 2000). Such collective approaches have been often referred to as eco-industrial parks (Côtè and Cohen-Rosenthal 1998), industrial symbiosis (Chertow 2004), industrial ecosystems (Côtè and Hall 1995), islands of sustainability (Wallner et al. 1996), industrial recycling networks (Schwarz and Steininger 1997), and by-product synergies (Forward and Mangan 1999).

What emerges from the literature is that IS approaches can be piloted from scratch or can occur spontaneously such as in the case of Kalundborg or the Humber district in the UK. Nonetheless once the spontaneous processes are initiated they are usually committee driven in order to be fully exploited. For example the activities in the Humber district are coordinated by the National Industrial Symbiosis Programme (NISP) network. Similarly the Kalundborg industrial district has its IS activities driven by the an association, constituted by the firms of the district, with a board of directors.

The application of IS methodologies can be driven by the economic benefit that can be attained by the symbiotic synergies between firms, or by the possibility of expanding a business, or by the fact that it may represent a means of reducing end of pipe burden of a productive system. IS may also be driven by state intervention in terms of public funding such as in the case of many US (Gibbs and Deutz 2004) and East Asian (Geng et al. 2009; Oh et al. 2003) IS projects.

Gibbs, Chertow and Mirata (Chertow 2007; Gibbs 2003; Mirata 2004) argue that the above mentioned state intervention, used to create from scratch IS interactions among firms, may generate cross organizational barriers that hinder the full potential of IS concepts. Moreover Chertow argues that self-organizing IS systems are likely to be more successful and develop to their fullest potential. Such successful development depends on the whether the productive systems already entail some basic existing symbiotic exchanges of which the actors may or may not be fully aware and it also depends on whether a coordinative organism (Schwarz and Steininger 1995) is at some point established in order to drive further the IS development. Baas (Baas and Boons 2004) state that for an IS system to emerge and evolve fully there must be a three stage development process. Firstly the firms within the industrial district must implement small synergic interactions among themselves to eliminate inefficiencies, such as for example utility sharing. This will then slowly get the firms to trust each other and it will bring about the exchange of knowledge and broaden their environmental awareness. Only then, once a certain amount of trust among the firms is established, can the interactions among the firms evolve toward more sustainable symbiotic exchanges of matter and energy.

As far as what is intended, in practical terms, by IS, by analyzing the literature concerning applicative IS scenarios, what emerged is that there can be various forms of joint industrial cooperation and symbiotic/synergic exchanges, namely:

- Symbiotic exchanges of waste to be used as by-products, energy (e.g., surplus energy or vapor) and water, (e.g., process water, refrigerant etc.)

- Joint management of infrastructures and services (e.g., roads, canteens, common spaces) and industrial systems (e.g., exhaust water treatment, energy production, waste disposal etc.)
- The symbiotic exchanges and the joint management can occur in the same firm, between groups of firms located in the same local geographical area and between groups of firms located within a large geographical area.
- IS is applicable among firms of the same or diverse economic sector, e.g., agriculture, fishing, forestry, industry, construction, commerce and services.

Based on the findings of the above mentioned literature study, the authors proceeded with analyzing who the potential IS actors in the Taranto province could be, identifying whether some IS was already present in the area. The next section details the methodologies and results of such analysis.

21.3 Economic and Environmental Analysis of the Industrial District

Since no current IS programs or initiatives were found to be active for the Taranto province, the first step of the work involved the study of the possible key players for a potential implementation of IS within the productive sector of the area. This entailed an economic and environmental analysis of the Taranto province productive sector.

For the economic analysis the sources of information used were the Italian National Statistics Institute (ISTAT) and the chamber of Commerce of the Taranto Province. The current population of the province is just over 570,000 and according to ISTAT 166,000 people are currently employed, most of which in the services sector (65 %) followed by industry (19 %), agriculture (11 %) and the building and construction sector (5 %).

By analyzing the Chamber of Commerce data, relative to the IV term 2010 and 2011, it was possible to identify the "active units" (firms with headquarters in the province) for each activity sector (according to the ISTAT economic activity classification ATECO 2007), which are 42,000.

The most important firms are: ILVA, a steel making plant with a maximum production capacity of 11.5 Mt/year of steel; the CET2 and CET3 thermal electric co-generation power stations, located in the ILVA site, with a total electric power of 1,065 MW that use exhaust gas to produce electricity and vapor; Cementir, a cement factory that uses blast furnace slag from the steelworks to produce cement with a maximum production capacity of 900,000 tons/year; the ENI refinery with a maximum production capacity of 6.5 Mt/year of refined crude oil; the area of the port of Taranto, the third biggest in Italy, composed of a military port a commercial port and an industrial port used mainly by the steelworks factory, the refinery and the cement factory; the VESTAS Centre that deals with the production and real time performance control of its wind turbines located in various wind power farms located in Puglia north Africa and Balkan area; Appia Energy, a power plant that manages and produces waste to create refuse derived fuel (RDF) for its electric energy power station; the Alenia Composite site (of the Finmeccanica Group) that produces sections of the fuselage Boeing 787 Dreamliner; the Dreher/Heineken brewery, one of the largest in Italy with a maximum beer production capacity of 200 ML/year. To complete the overall picture, when considering the provincial industrial firms registered as active units, 23 % of these operate in the food and drink sector, 20 % in the metal products sector, 10 % in the clothing sector, 10 % in the non-metal bearing mineral products sector and 7 % in the wood products sector.

Having determined the most important firms and industrial sub-sectors in economic terms, an environmental mapping of the industrial sector was performed. The information for the this mapping was sourced from environmental studies and reports of different firms, from the Taranto Chamber of Commerce, from internet and specific firm related websites and from a report for the area of Taranto (commissioned by Ministry for the Environment, written by the ENEA in 1996) regarding the identification of environmental and safety issues for areas with a high risk of environmental crisis.

As expected, from the study of the above mentioned sources, the most impacting industrial firms in the province are the large ones that were previously identified during the economic mapping of the area.

The main emissions and impacts of these large industrial complexes are typical of industrial areas (e.g., CO, CO_2, sulfur oxides, Nitrogen oxides, dust, noise) with the addition of benzo[a]pyrene, heavy metals, dioxins and volatile organic compounds which are distinctive emissions of steelworks plants and refineries. The large industrial complexes are also responsible for a large land use, for example the steelworks and refinery sites occupy nearly 1,800 ha of land next to the city of Taranto. Similarly there is also a large amount of water use mainly ascribable to the beer factory, the refinery and the steelworks. The port activities, together with the other complexes in the Taranto area, are mainly responsible for the intense road traffic throughout the province.

21.4 Waste and Energy Analysis

The waste analysis was performed for the whole industrial sector of the province, starting from the firms identified in the previous phase of the project, as a means of identifying the main quantities of waste produced, disposed and recovered. Information on such waste flows is necessary for the recognition of possible better waste uses and exchanges among firms, at a symbiotic/synergic level, in order to improve the sustainability of the industrial sector.

The data regarding waste generated and managed by firms in the province was sourced from the Chamber of Commerce and from Ecocerved, a company of the Chambers of Commerce consortium that operates in the area of environmental IT.

The most recent available data regarded the year 2008. By analyzing such provincial information what emerges is a yearly production of 3.17 Mt of waste, whilst 3.99 Mt are recovered or disposed during the same year. This implies that nearly a million tons of waste is originating from outside of the province. Of the waste produced, over 53 % is produced from thermal processing and one quarter is due to construction and demolition activities. The waste recovered, approximately 3 Mt/year, mostly originates from thermal processing (50 %) and from the construction and demolition activities (29 %). However 90 % of the recovery includes activities such as composting, land filling, spreading on ground and stocking for future recovery. Such activities could be substituted with others that involve more efficient and useful waste recycling. Over 70 % of the total waste disposed (1 million t/year) is deriving from water treatments.

A detailed analysis of the single firms responsible for the production, recovery and disposal of at least 1,000 t/year indicated that 45 firms are responsible for 38 types of waste in such quantities. Some of the most interesting types of waste, in terms of quantities per year are:

- Untreated blasts furnace slag—1.5 Mt—produced by the steelworks. Only 15 % of this is used by the local cement factories, the rest is shipped to south America.
- Untreated basic oxygen furnace slag—1.5 Mt—produced by the steel works. All used for land filling within the steelworks site.
- Refractory and other coating materials from metallurgy 87 kt—mostly produced by the steelworks. Principally disposed of in dumps;
- Mill scale—currently 70 kt produced by the steelworks ILVA are disposed.
- Coal fly ash 38 kt—Imported from outside the province;
- Leachate from dumps—61 kt produced, 14 kt disposed in specific dumps;
- Sludge from water filtration processes—produced by various firms, 57 kt disposed in dumps.

What emerges is that there are a large quantity of certain waste types, originating mainly form the steelworks and a few other companies, that could be possibly reused in a more efficient manner compared to current practices. Section 21.6 illustrates some possible re-utilizations of such waste.

As far as energy is concerned, the analysis for the province was performed using data from the 2006 Regional Energy Plan, the 2008 Regional Energy Balance report drafted by the national state agency ENEA and 2010 data from TERNA, the company responsible for the national electricity grid. By analyzing the data regarding energy transformed into secondary sources, what emerges is a total 7,119 ktoe (tons oil equivalent) was transformed in 2010. Of this total 89 % is due to the refining of petroleum and the rest is mainly due to the burning of natural gas in the steelworks power stations.

By looking at the data regarding the consumption per sector, what emerges is that the industrial sector is responsible for 87 % of the energy consumption whilst other sectors (residential, tertiary, agriculture and transport) are responsible for the rest. Of the total energy utilized by the industrial sector (4,353 ktoe), 85 % is used by the steelworks, 11 % by the refinery and 2 % by the cement factory. Seventy-eight percent of the total energy consumption for all sectors is due to the use of coal which is solely used by the steelworks plant. When considering the transformation losses in terms of cooling water, a total of 1,018 ktoe is sent out to sea without any heat recovery from the main power stations in the province. Of this total 88 % derives from the steelworks power stations. If compared to the total energy used in the provincial residential sector (169 ktoe), the above mentioned transformation energy loss represents a vast amount of energy, currently wasted, that could be usefully recovered and reutilized. Of the total consumption of electrical energy in the province 77 % (4,720 GWh) is consumed by the industrial sector. Again the steelworks sector dominates being responsible for 81 % of the sectors' consumption, followed by the refinery (10 %), the cement factory (2 %) and the construction materials industry (2 %). The rest of the industrial sector, constituted mainly of medium and small firms, accounts for 5 % of the total industrial electricity consumption. Similarly to the waste analysis results in the Sect. 21.3, the steelworks dominates, in energetic terms, over the whole province.

21.5 Current Levels of IS in the Province

In order to gather extra IS information regarding the provincial industrial sector, over 500 questionnaires were sent out to industrial firms. Unfortunately there was very little response from such firms, but nonetheless the previously described phases of the project allowed the assessment of the current state of symbiotic/synergic interactions present in the province. The cement factory reutilizes 15 % of the total blast furnace slag produced by the steelworks plant. The power plants within the steelworks factory make use of the exhaust gases to produce electricity. The refinery, via its power plant, produces electrical energy together with demineralized water and vapor from sea water and exchanges it with the ILVA steelworks. Metal scraps collected from around the province are used by the steelworks during the steelmaking process. Waste management and disposal firms input RDF to the Appia Energy waste-to-energy power plant to produce electricity. Coal fly-ash waste (from outside the province) is used by construction material producing companies and cement factories to produce building material and cement. Finally in the agricultural sector, distilleries use the marc from the wine companies to produce high alcohol content beverages.

21.6 Some New Possible Symbiotic Interactions Within the Province

What follows are some indications of new possible symbiotic interactions that could be achieved in the province based on the findings described in the previous sections. In particular the main features of possible waste reutilization and better energy management schemes are described in the following paragraphs. A detailed description of these approaches can be found in (Notarnicola et al. 2012).

For each of the types of waste identified in Sect. 21.4 some alternative uses, found in literature (publications, internet, books, journals etc.), as raw materials for other industrial production, are listed below. These are by no means exhaustive and need to be verified in detail in order to ascertain their effective exploitation. Nonetheless these alternative reutilizations represent a starting point for a future study for a real implementation of IS concepts in the province.

The 1.5 Mt/year of blast furnace slag, generated by the steelworks plant, 85 % of which is exported to south America, could be used to produce asphalt based paving, marine embankment containment, bricks, mineral wool, improvement of soil structure, glass and fertilizers. The 1.5 Mt/year of basic oxygen furnace slag, generated by the steelworks and currently used for land filling, could be used to produce cement, road construction material and systems for phosphorous removal from waste water. The 70 kt/year of mill scale, generated by the steelworks and disposed, could be used to produce sintering briquettes and pellets for steelworks, alluminothermic welding, casting sands, electric arc electrodes, pigment for paints/cements/cosmetics, ammonia (mill scale used as a catalyst), cement, cement for radiation shielding, glass, refractory materials, iron salts, batteries, hydrogen, improvement of soil structure, water treatment systems and hydrogen fuel cells. The 78 kt/year of refractory materials, generated by the steelworks and disposed, could be used to produce cement, bricks, plaster, embankments, improvement of structure quality of soil and electric arc furnace steel. The 38 kt/year of fly ash, imported into the province and currently used to produce mainly cement, could be used to produce hydraulic lime, glass, ceramics, improvement of structure quality of soil, road paving and fertilizers. The 57 kt/year of water sludge from sewage treatment and filtration, currently disposed, could be used to produce bricks and fertilizers. Finally the 15 kt of leachate from dumps currently disposed in special dumps could be used to produce hydrogen and fertilizers.

Referring to the energy used, the over 1,000 ktoe of thermal energy (Sect. 21.4) wasted at sea by the industrial sector could represents a great opportunity for heat recovery and reutilization. Some useful ways of recuperating this wasted energy are:

- Organic Rankine Cycle (ORC) systems used to produce electricity from heat
- Co-generation of useful heat and electricity in a combined heat and power plant
- District heating: the industrial waste water could be passed in a heat exchanger. The heat could be then transferred to a piping system used to deliver hot water to domestic housing.
- Use a heat exchanger to pre-heat the liquid or the gas of heat pumps, of absorption heat/cooling transformers or the fuel of burners.

Such energy recovery approaches, that could be implemented by one or more firms in conjunction with others, could help improve the sustainability of the industrial sector reducing the overall energy consumption. Other similar approaches include the use of more efficient electric motors and compressors as well as improving the overall energetic performance of industrial buildings.

21.7 Conclusions

The study entailed in this chapter has highlighted various aspects of the possibility of implementing IS concepts to the industrial district of the Taranto province. The economic, environmental, waste and energy analysis of the provincial industrial sector, has identified who the key players of a IS scenario in the province could be. Moreover the study has identified some of the currently existing IS interactions in the province, especially among the larger industrial complexes. The existence of such interactions, as pointed out by Chertow (Chertow 2007), is a positive factor since it is more likely that further IS interactions can be developed fully. However to develop such interactions an organizing committee is desirable to guide the flow of information among the IS actors. Such committee must have an organizational side that effectively gathers the information and makes it available to the correct people involved. There must also be a scientific part to the committee that gathers the scientific information that should be used to stimulate the development of synergic interactions. The universities can certainly have a central role in this scientific side of the committee. Initially the role of the future committee, as pointed out by Baas (Baas and Boons 2004) will be that of getting the IS role players to know each other and to share information among themselves in order to build up the reciprocal trust. This should then be followed by the development of small synergies among firms such as for example utility sharing. Only when there has been an overall build-up of trust among the interacting firms can there be a shift to matter and energy exchanges within the province. As a starting point for these kind of exchanges, the ones suggested in this chapter could be used. In fact, the new possible exchanges, described in the previous sections, make use of the major identified quantities of inefficiently used and disposed waste and by products (over 3.25 Mt) and waste-energy (over 1,000 ktoe of heat), mainly originating from the steelworks plant. Since these reported exchanges are non-exhaustive the above mentioned future committee will have to study them further (in technical, environmental, financial-economic and legal terms) and try, if possible, to propose them to the firms for a real implementation of IS interactions that will effectively be able to make the Taranto province industrial system more competitive and environmentally sustainable.

References

Baas LW, Boons FA (2004) An industrial ecology project in practice: exploring the boundaries of decision-making levels in regional industrial systems. J Clean Prod 12(8–10):1073–1085

Boons F, Spekkink W, Mouzakitis Y (2011) The dynamics of industrial symbiosis: a proposal for a conceptual framework based upon a comprehensive literature review. J Clean Prod 19:905–911

Chertow MR (2000) Industrial symbiosis: literature and taxonomy. Ann Rev Energy Environ 25:313–337

Chertow M (2004) Industrial symbiosis. In: Cleveland CJ (ed) Encyclopedia of energy. Elsevier, Oxford

Chertow MR (2007) Uncovering industrial symbiosis. J Ind Ecol 11(1):11–30

Côtè RP, Hall J (1995) Industrial parks as ecosystems. J Clean Prod 3(1–2):41–46
Côtè RP, Cohen-Rosenthal E (1998) Designing eco-industrial parks: a synthesis of some experiences. J Clean Prod 6:181–188
Cutaia L, Barberio G, Li Rosi O, Mancuso E (2011) Verso una piattaforma di simbiosi industriale: l'attività in corso nell'ambito del progetto Eco-innovazione Sicilia. In: Proceedings of Ecomondo 2011, Rimini, Italy
Dinelli D, Garro S, Giovannelli L, Querci P (2003) Progetto CLOSED: il modello DPSIR applicato ai distretti di Prato, Lucca, Pistoia. ARPAT-Toscana, Firenze
Forward G, Mangan A (1999) By-products synergy. Bridge 29(1):12–16
Geng Y, Zhang P, Coté R, Fujita T (2009) Assessment of the national eco-industrial park standard for promoting industrial symbiosis in China. J Ind Ecol 13(1):15–26
Gibbs DC (2003) Trust and networking in interfirm relations: the case of eco-industrial development. Local Econ 18(3):222–236
Gibbs D, Deutz P (2004) Implementing industrial ecology? Planning for eco-industrial parks in the USA. Geoforum 36(4):429–439
Mirata M (2004) Experiences from early stages of a national industrial symbiosis programme in the UK: determinants and coordination challenges. J Clean Prod 12(8–10):967–983
Notarnicola B, Uricchio AF, Tassielli G, Renzulli PA, Selicato G (2012) Elaborazione di un modello di applicazione dei principi e degli strumenti dell'ecologia industrial ad un'area vasta. Cacucci, Bari
Oh D, Kim KB, Jeong SY (2003) Eco-industrial park design: a Daedeok Technovalley case study. Habitat Int 29(2):269–284
Romano M, Raggi A, Morgante A, Tascione V (2010) Industrial symbiosis and eco-industrial parks: a preliminary study of potential scenarios in the Sangro-Aventino industrial area. In: Proceedings of the 1st International Conference small-business Industrial Clusters in China and Italy. Chieti-Pescara, Italy, May 28–29 2010
Schwarz EJ, Steininger KW (1995) The industrial recycling-network enhancing regional development. Research Memorandum #9501. Department of Economics, University of Graz, Austria
Schwartz EJ, Steininger KW (1997) Implementing nature's lesson: the industrial recycling network enhancing regional development. J Clean Prod 5(1–2):47–56
Wallner HP, Narodoslawsky M, Moser F (1996) Islands of sustainability: a bottom-up approach towards sustainable development. Environ Plann A28(10):1763–1778

Chapter 22
Harmonization of Documentation for EPD-Generation: The Norwegian Model

Annik Magerholm Fet, Marina Magerøy, Dina Margrethe Aspen and Dagfinn Malnes

Abstract This chapter gives an overview of the different requirements that have to be fulfilled for the generation of an EPD. The Norwegian EPD Foundation (in brief *EPD-Norway*) is the Norwegian organization for verification and registrations of EPDs for different types of products. EPD-Norway was established in 2002 and since then 23 PCRs and more than 200 EPDs have been registered under this system. However, analyses have shown that the format of the PCRs and EPDs within the Norwegian system have differed, and compared with international systems even greater differences have been observed. To compensate for this difference EPD-Norway has put effort into a harmonization of PCR-templates, into LCA-reporting practice and to EPD-schemes. A set of underlying procedures and databases are further developed to guide companies through the process of generating EPDs. The chapter will present the results of this process and ideas for further development of the system.

Keywords Environmental product declaration (EPD) · Product category rules (PCR) · Life cycle assessment (LCA) · PCR-templates · EPD-Norway

22.1 Introduction

In today's globalised world, focus on sustainable development and corporate social responsibility in business is crucial for progress and prosperity. During the past 50 years the perspective on environmental issues and the approach to which it has been dealt with has changed significantly. Figure 22.1 shows the main strategies used for every decade since the 1960s. This will be further elaborated in the following, mainly focusing on environmental issues and the measures used to solve these issues.

A. M. Fet (✉) · M. Magerøy · D. M. Aspen
Department of Industrial Economics and Technology Management, Norwegian University of Science and Technology, NTNU, Trondheim, Norway
e-mail: annik.fet@iot.ntnu.no

D. Malnes
The Norwegian EPD Foundation
P.O. Box 5250 Majorstuen, 0303 Oslo, Norway

R. Salomone, G. Saija (eds.), *Pathways to Environmental Sustainability*,
DOI 10.1007/978-3-319-03826-1_22, © Springer International Publishing Switzerland 2014

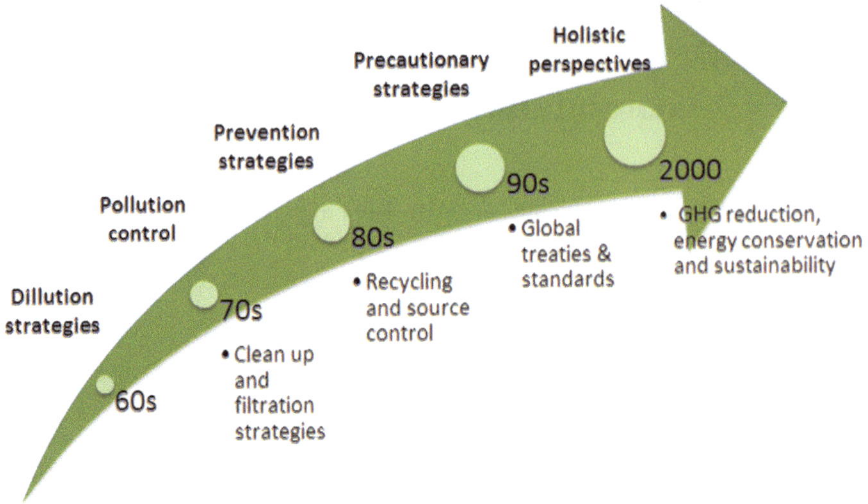

Fig. 22.1 Overview over development of approach towards environmental strategies

Environmental management systems, environmental auditing and reporting, and documentation of impacts are important tools in use. Such tools can be applied at all levels, the products life cycle, corporate site and the global supply chain. Environmental product declarations (EPD) are used to communicate life cycle based data regarding the environmental profile of products and services, and can be used as a tool in environmental management. The main purpose of EPDs is to provide quantified measure of the environmental impacts of a product or service to professional purchasers, management, government and consumers. Important characteristics of EPDs are objectivity, comparability and validity.

In 2006 ISO 14025—Environmental Declarations Type III (ISO 2006a) was published, an international standard with principles and procedures for the development of PCRs and EPDs. The process towards the ISO 14025 standard was long, and the work in ISO had been going on since 1994. The idea of environmental product declarations was first presented to the members of the ISO Technical Committee 207, Sub Committee 3 (TC207/SC3) by an international expert of the American delegation (Bogeskär et al. 2002). The issue was put on hold till the meeting in 1995 where the Swedish delegation lobbied to put EPD back on the agenda. During the next two years Working Group 1 with the support of a special task group, started working on the creation of an international standard. In 1998 the standardization work was interrupted to create a Technical Report Type 2 which was published in 1999. The Technical Report summarized the discussions in ISO and presented the state of the art on EPD. After further development the standard was finalized and published in 2006.

22.2 The History Behind EPD-Norway

The development and use of EPDs had been going on for some time prior to the standard, and several national EPD programs have developed both before and after the standard was published. EPD Norway was established in 2002 by the Confederation of Norwegian Enterprise (NHO) and the Federation of Norwegian Building Industries (BNL). The reason for its establishment was an expressed desire for standardized and internationally valid EPDs for products and services. EPD Norway has the task of verify and register the EPD according to the international standard ISO 14025. According to this standard, the EPD has to be verified by an independent third party.

It is important to the corporate sector that all environmental information be provided in an objective, standardized fashion, and that the competitive conditions are equal for both Norwegian and foreign enterprises. The goal of EPD Norway to make EPD as the preferred communications tool when exchanging environmental information about products and services across all sectors both nationally and internationally.

EPD Norway collaborates with the leading international research institute SINTEF Building and Infrastructure, Ostfold Research Co. and the Norwegian University of Science and Technology (NTNU) among others to develop new Product Category Rules (PCR). Together with industry organizations EPD Norway gives financial support to the development of product category rules.

EPD Norway Foundation is responsible for ensuring that Norwegian PCRs comply with the requirements in the ISO 14025 standard as well as ensuring that EPD's are developed in accordance with the sector-specific PCR.

22.3 Overview of Programs and Existing Requirements to LCA, PCR and EPD

Through Life Cycle Analysis (LCA) the impacts of a product or service is calculated over its entire lifetime, from cradle to grave. This includes extraction of raw materials, production/manufacturing, transportation, use face and end-of-life phase (waste handling or recycling). By looking at the entire production system, the LCA can help avoid shifting environmental burdens from one stage to another and sub-optimization (Baumann and Tillman 2004). Economic and social aspects are not included in LCA. LCA refers to both the results of the analysis and the activity of establishing these results (Heijungs 2007). Life Cycle Analysis forms the basis of an EPD, and should be performed according to ISO 14040:2006 "Environmental Management Life Cycle Assessment Principles and Framework". In addition LCA can be used in environmental management systems to identify significant environmental aspects of products and services and to integrate environmental aspects into product design and development. Other areas of use include environmental communication, environmental accounting and environmental impact assessments (ISO 2006b).

LCA is seen as an impartial analysis and is the basis of environmental policy in the EU, Japan and elsewhere in the world (Schenck 2010). An LCA study consists of four phases (ISO 2006b):

- Goal and scope definition,
- Inventory analysis,
- Impact assessment and,
- Interpretation.

The product of study and the purpose of the LCA are specified through the goal and scope. The inventory analysis (LCI) is the phase where data collection and calculation procedures are defined. The impact assessment relates the emissions and resources to environmental problems through classification and characterization. Finally the interpretation considers the results from the analysis and impact assessment together and in light of the goal and scope of the analysis. The results are then presented, and can in some cases take the form as conclusions or recommendations for decision-makers (ISO 2006b).

An EPD provides an impartial and neutral quantified presentation of the environmental impacts of a product or service, based on a life cycle analysis. EPDs are developed on the basis of PCR. PCRs are in general developed by an EPD-program operator in cooperation with LCA- and industry experts, and in practice they are guiding documents to be used when developing EPDs. In addition to requirements of format and content of the EPD, the PCR states the scope and goal of the LCA to be performed. Important factors of the PCR are life cycle stages to be included, parameters to be covered, and the way in which parameters are collated and reported (Fet and Skaar 2006). Only one PCR is developed for every product category, and the harmonization of existing PCRs across EPD programs is encouraged. If an existing PCR can be used, it is to be preferred (ISO 2006a). The requirements to the content of a PCR are presented in Fig. 22.2.

EPD are primarily aimed at being used in 'business-to-business'- communication, but can also be used by interested consumers. The information attained from EPDs can be used for various activities, and EPDs can be said to have five different functions (Hillier et al. 2004):

1. Management tool
2. Communication tool
3. Evaluation/assessment tool
4. Political tool
5. Action tool

In management the information from the EPD can be used in environmental management, product innovation and design (Kristensen et al. 2006) to improve the environmental performance of the product and for purchasers to give the environmental profile of the product.

No.	Criterion
1	Definition and description of product category
2	Functional or declared unit
3	System boundaries
4	Description of data
5	Cut-off rules
6	Data quality requirements
7	Units
8	Inventory analysis
9	Impact category selection & calculation rules
10	Predetermined parameters for reporting LCA data
11	Requirements for additional environmental information
12	Materials and substances to be declared
13	Format and content instructions
14	Period of validity

Fig. 22.2 Requirements of contents of PCR from ISO 14025 shown as criterion for compliance. (ISO 2006a)

The EPDs offer environmental information which can be communicated throughout the supply chain and to purchasers and other customers, in addition to communicating environmental awareness. EPDs are used by purchasers and customers to evaluate and compare different products through benchmarking. The political aspect of EPDs includes providing government the tool to disseminate environmental consciousness in the business and public sector (Hillier et al. 2004). In addition EPD can be used internally to encourage improvement of the environmental performance of a product (ISO 2006a). Through the information provided in an EPD the company is able to identify hotspots in the production chain where improvements can and should be made. An EPD is not a static document, and can be continuously updated after it is published. In the building and construction sector EPD are used to assess environmental impact of entire buildings.

The development of an EPD is a very comprehensive process of data collection, and difficulties can be faced in achieving quantitative data from suppliers (Kristensen et al. 2006). Additionally this can be a cost intensive process for the developer.

To ensure credibility of the EPD it is important that the program operator establishes procedures for the verification of the LCA-data, PCR and EPD. Credibility can also be enhanced through transparency. It is therefore important for interested parties to have access to information through all the stages of development and operation of EPD.

Today EPD development is done through national EPD-programs. Some of the most developed EPD-programs found globally are presented in Fig. 22.3 which shows a summary of the main features of each EPD program (Magerøy 2011). For fields that have been left blank, information has not been available. As can be seen there is a variance between the types of scheme owners. Some programs are owned by governmental organizations, such as Sweden, South Korea and France. The remaining programs are owned by various private organizations from the industry, research institutes or foundations which are a mixture of both.

Country	Programme name	Scheme owner	Main focus	From	Standard
Norway	Norwegian EPD Programme	Confederation of Norwegian Enterprise	Building products and furniture	2002	ISO 14025, 21930
Sweden	International EPD System	Swedish Environment Management Council	Various products	1997	ISO 14025, 21930
Finland	RT Environmental Declaration	Building Information Foundation, RTS Confederation of Finnish Construction Industries RT	Building Products	1988	ISO 21930
Germany	IBU	Institute of Construction & Environment	Building Products	1998	ISO 14025, 21930
Japan	Eco-Leaf	Japan Environmental management for Industry	Electronics	2002	ISO 14025
South Korea	Type III Labelling programme	Ministry of Environment	Electronics	2001	ISO 14025
Denmark	Miljøvare-deklarationen	Fonden Dansk Standard		2009	ISO 14025
France	INIES	Département Energie, Santé et Environnement	Building Products		NF P01-010
Netherlands	MRPI	Industry	Building Products	1997	ISO 21930
United States	Earthsure	Institute for Environmental research & education	Food & Agriculture	2000	ISO 14025
United States	Eco-Profile Programme	Scientific Certification Schemes			ANSI LCA standard

Fig. 22.3 Summary of EPD programs. (Magerøy 2011)

22.4 How Has This Been Harmonized Through the Work in EPD-Norway

One of the main purposes of an EPD is to compare the environmental impacts of similar products or services. In order for two EPD to be comparable, it is imperative that the requirements for developing the EPD are identical or equivalent, as given in ISO 14025. If two EPD are based on the same PCR, the EPD will be comparable if all the requirements of the PCR are followed, and the PCR includes all the requirements of the standard. Even though the standard has existed for some years now, there are still great variances between the development, content and presentation of EPDs both between, and within the programs (Magerøy 2011). Harmonization between programs and its guidelines, and interpretation of the standard is necessary in order for the three characteristics of EPDs to be fulfilled.

The Norwegian EPD Foundation has an on-going project with the goal of striving for comparability between EPDs. The main areas of focus of their work is on the development of new EPDs, and thereby development of PCR and EPD guidelines, in addition to recommendations on EPD contents and layout. Through more clear and harmonized guidelines the goal is that Norwegian EPDs will be more uniform and thereby increase comparability. Therefore, during the past few years, EPD Norway has done stepwise projects with the goal of harmonizing the structures, formats and contents of EPD-related documents. As stated in ISO 14025, the program operators shall attempt to use existing PCR and standards whenever possible when new product categories are integrated in the program, and work to harmonize relevant documents and information pertaining to EPD generation. This has resulted in three processes where templates for PCRs, EPDs and LCA reports have successively been created and revised in the period 2007–2012.

22.5 The Harmonization Process

Several initiatives were initiated by The Norwegian EPD Foundation with the goal to achieve standardized formats of a PCR, an EPD and an LCA-report manual. A group of experts was established with the mandate to organize workshops, meetings, and hearings of draft-documents and testing of these until a set of documents was approved. The Norwegian EPD Foundation is organized in a small organization with a director (full time employment), one verification officer in 50% employments and a secretary in 50% employment. However, much of the work is organized through committees (Technical committee, verification committee and communication committee) where experts from different companies, research institutions and universities are working on a voluntary basis. The harmonization project has been carried out as a project with a small budget, but with voluntaries from the different organizations contributing with their expertise in various ways.

22.5.1 Project Organization

The project was anchored in the technical committee, but the project was performed as collaboration between experts from NTNU, Sintef and Ostfold research (Fet et al. 2011b) and the work tasks were organized in the following way:

- Preparation of PCR-model (lead by Ostfold research)
- Preparation of EPD-model, short and long version (lead by NTNU)
- Preparation of LCA-report structure (lead by NTNU)
- Preparation of guideline for how to develop EPDs (lead by Sintef Byggforsk)

Each expert group prepared their drafts and the respective underlying documentation, which were further presented at a workshop in spring 2011.

22.5.2 Workshop Spring 2011

The goal of this workshop was to prepare for generation of EPDs that secure comparison of information on products with the same functions. The focus on this workshop was to evaluate the draft documents and their layouts. It was further a discussion around how to group the PCRs (according to the CPC-system or other), and if the EPDs should contain information in addition to the main requirements in ISO 14025, e.g. if information on Carbon footprint, chemicals and health impacts should be included in the EPDs or not. In addition the content of the guideline and how detailed the information about methodological decisions, datasets and the difference B2B and B2C, should be presented in the guideline. Also, guidelines on which electricity mixture to use in the datasets, i.e. the proportions of various energy carriers such as fossil, hydropower, nuclear, biomass, solar, wind and the like, was discussed. A further discussion around the harmonization to other EPD-programs was a topic for discussion. The outcome from the workshop discussion was documented in a separated report for background information.

22.5.3 Draft Document on a Hearing

Based upon the draft documents and the discussions in the workshop, revised versions of the documents were sent on a hearing which was organized by the technical committee in The Norwegian EPD Foundation.

22.5.4 Reporting

A report was prepared for the TC207-meeting in Oslo, June 2011. This report summarized the document from the hearing and the guideline for development of PCRs, EPDs and LCA-reports. The report further outlined the plan for how to take the documents further to approved models for the Norwegian EPD-work (Fet et al. 2011b).

The report delivered in 2011 has laid the foundation for further discussions and document development, especially regarding:

- Content: How to determine on what to harmonize against (use of reference documents and other reference information, how to address differences between countries, industries etc.)
- Formats: How to create documents that are effective at communicating requirements and guidelines, and ensure efficient use and dissemination

A larger portion of the EPDs registered by The Norwegian EPD Foundation, is on products of relevance for the construction sector. Since the research institute SINTEF Building and Infrastructure sector has been one of the organizations that have contributed to the establishment of The Norwegian EPD Foundation, the models developed for PCRs and EPDs in Norway have therefore been influenced by this sector. The standard PCR- and EPD-documents are therefore in compliance both with ISO 14025 and EN 15804 (CEN 2012) which is European standard for EPDs and PCRs for construction products.

22.6 The Harmonized Documents

22.6.1 The PCR for the Norwegian EPD Foundation

Steps to carry out before the creation of a PCR are; firstly, to determine the product category to which the PCR will be applicable; secondly, to identify a program operator; thirdly, to perform a search for PCRs that belong to the same category. If a PCR for the actual product is not in existence, then a PCR should be developed according to the PCR-guide as shown in Fig. 22.4. The main content of the PCR-guide approved by The Norwegian EPD Foundation is:

1 Introduction
2 Resumé and cross references
3 Scope
4 Terms and definitions
4.1 Environmental product declaration (EPD)
4.2 Life cycle assessment (LCA)
4.3 Declared Unit
4.4 Functional Unit
4.5 <Product group> specific definitions
5 General aspects
5.1 Types of EPD
5.2 Comparability of EPD of construction products
5.3 Communication format
6 Product Category Rules for LCA
6.1 Product Category
6.2 Life cycle stages and their information modules to be declared
6.2.1 General
6.2.2 A1-A3, Product stage, information modules
6.2.3 A4-A5, Construction process stage, information modules
6.2.4 B1-B5, Use stage, information modules related to the building fabric
6.2.5 B6-B7, use stage, information modules related to the operation of the building
6.2.6 C1-C4 End-of-life stage, information modules
6.2.7 D, Benefits and loads beyond the system boundary, information
6.3 Calculation rules for the LCA
6.3.1 Functional unit
6.3.2 Declared unit
6.3.3 Reference service life (RSL)
6.3.4 System boundaries
6.3.5 Criteria for the inclusion of inputs and outputs (cut-off)
6.3.6 Selection of data
6.3.7 Data quality requirements
6.3.8 Scenarios on product level
6.4 Inventory analysis

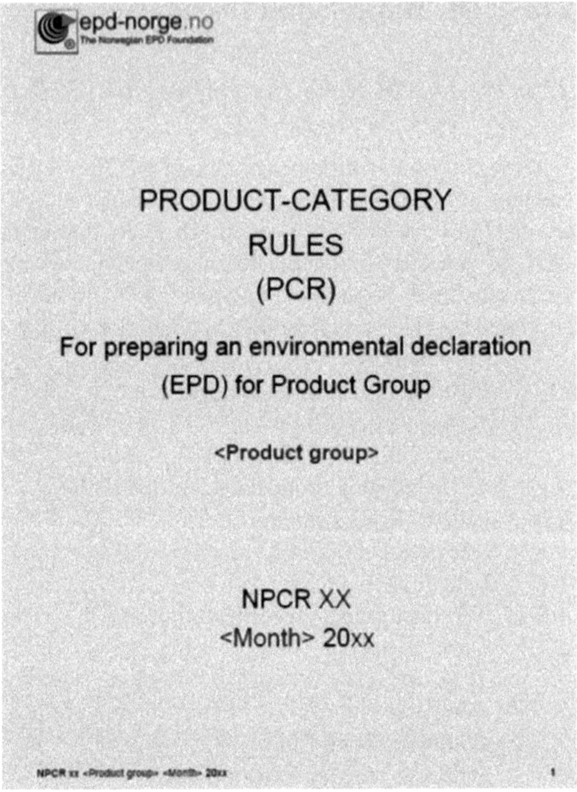
epd-norge.no
The Norwegian EPD Foundation

PRODUCT-CATEGORY
RULES
(PCR)

For preparing an environmental declaration
(EPD) for Product Group

<Product group>

NPCR XX
<Month> 20xx

NPCR xx <Product group> <Month> 20xx 1

Fig. 22.4 PCR-guide for the Norwegian EPD Foundation

6.4.1 Allocation of input flows and output emissions
6.5 Impact assessment
7 Content of the EPD
8 Project report
9 Verification and validity of an EPD
10 References

If a PCR exists, but it is not possible to develop a unified PCR or adapt the existing PCR, the program operator should look for possibilities to align with the existing PCR. Alignment refers to the process of making PCRs consistent. Alignment needs to occur across three principle elements for complete alignment to be realized: (1) data; (2) the LCA and additional information rules of the PCR; and (3) PCR-related procedures.

22.6.2 *The EPD for the Norwegian EPD Foundation*

Most of the EPDs developed in Norway have followed the EPD-format based upon the Nordic project on implementation of EPDs, the so-called NIMBUS-project

No.	Requirement
1	Organisation identification and description
2	Product description
3	Product identification
4	Programme information
5	PCR identification
6	Date of publication and period of validity
7	LCA/LCI data
8	Information about life cycle boundaries
9	Additional environmental information
10	Content declaration
11	Mandatory statement
12	Reference to explanatory material
13	Demonstration of verification

Fig. 22.5 Criterion for contents of EPD from ISO 14025

(Hansen et al. 2001). They are also meant to be in compliance with the ISO 14025, see requirements given in Fig 22.5.

However, an analysis performed in 2011 (Magerøy 2011) shows that among ten EPDs chosen within each of the five product categories of the Norwegian EPD-program, none of the assessed EPDs fulfill all requirements of ISO 14025, see Fig. 22.6.

The EPDs have to some extent the same format and contents, but depending on the organization producing the EPD, there are variances within formulations, which information that is included and with regards to color and logo use.

The requirement that most EPDs do not fully meet is the organization identification and description. All EPD include an organization identification consisting of name, logo and contact information, but only three of the assessed EPD include a short description of the producer. In addition three EPDs lack a description of the product. Three EPDs do not have the date of publication included in the EPD, only the date of which it is valid to. Two EPDs do not have a content declaration of the materials, both of these from the product category "Chemicals". Instead, the EPDs include a more detailed description of the product and its production process. Three EPDs, all of which are prepared by different organizations, does not include the mandatory statement. One EPD lacks reference to explanatory material. Three EPDs, all of which are prepared by different organizations, does not include the mandatory statement. One EPD lacks reference to explanatory material.

So, with this considerable variation, the technical committee has focused on an EPD-format that will meet all requirements in ISO 14025 and in EN 15804. The first step was to create a first page that should look similar on all EPDs registered by the Norwegian EPD Foundation, see Fig. 22.7.

The step further is to establish a common format for EPDs developed from different standards, ISO 14025, ISO 21930 (ISO 2007) and EN 15804. The heading from the front page is shown in Fig. 22.8, and will be introduced early in 2013. A program for transforming existing EPDs into the new format will also be introduced at the same time.

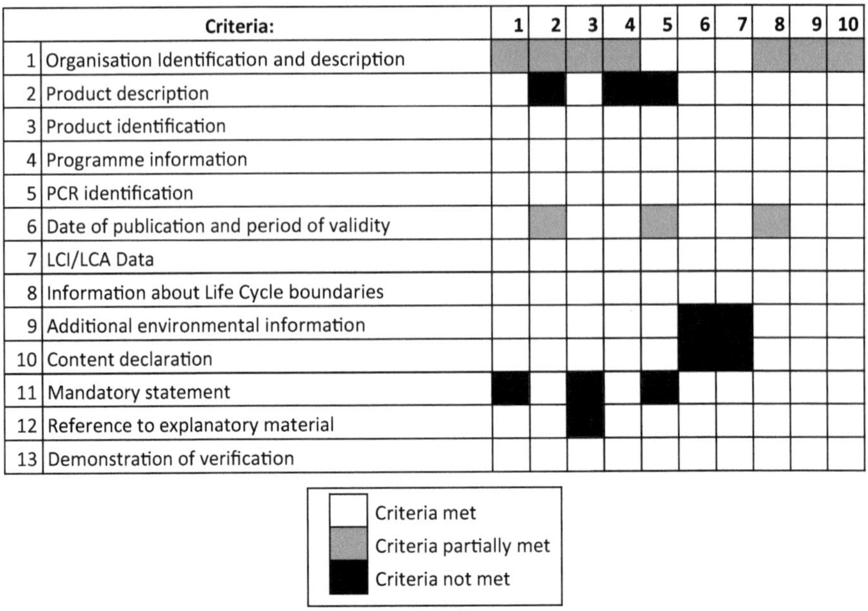

Fig. 22.6 PCR-guide for the Norwegian EPD Foundation

22.6.3 The LCA-Reporting Format for the Norwegian EPD Foundation

It is a requirement of ISO 14025 that the PCR be based on "one or more life cycle assessments (in accordance with the ISO 14040 series of standards)". It is ideal if a LCA is meant as the basis of the PCR, to ensure that the scope, LCI, and LCIA phases all match. While all LCAs should be conformance with ISO 14044 (ISO 2006c), it is also ideal if it is certified by an external third party. The LCA should be current, representing the current product design, market condition, and material sourcing, all based on the product's lifetime in the market. A newly developed product should have a new LCA. The functional unit used in the LCA must be directly applicable to the PCR.

To achieve sufficient quality of the LCA-reports developed by different LCA-experts or by companies themselves, The Norwegian EPD Foundation support a project in 2012 with the goal of harmonizing the LCA-report format. One important purpose of this project was to help verifiers to control that the LCA has been performed and that the verifiers could rely on the project has been performed in accordance with the ISO 14044:2006, ISO 14025:2006 and the actual PCR documents. An LCA-report format will help to communicate the expectations to the reporter and contribute to more standardized LCA-reports and verification processes. A guidance for LCA-reporting can also include recommendations from the technical committee relative to the other existing recommendations and guiding documents.

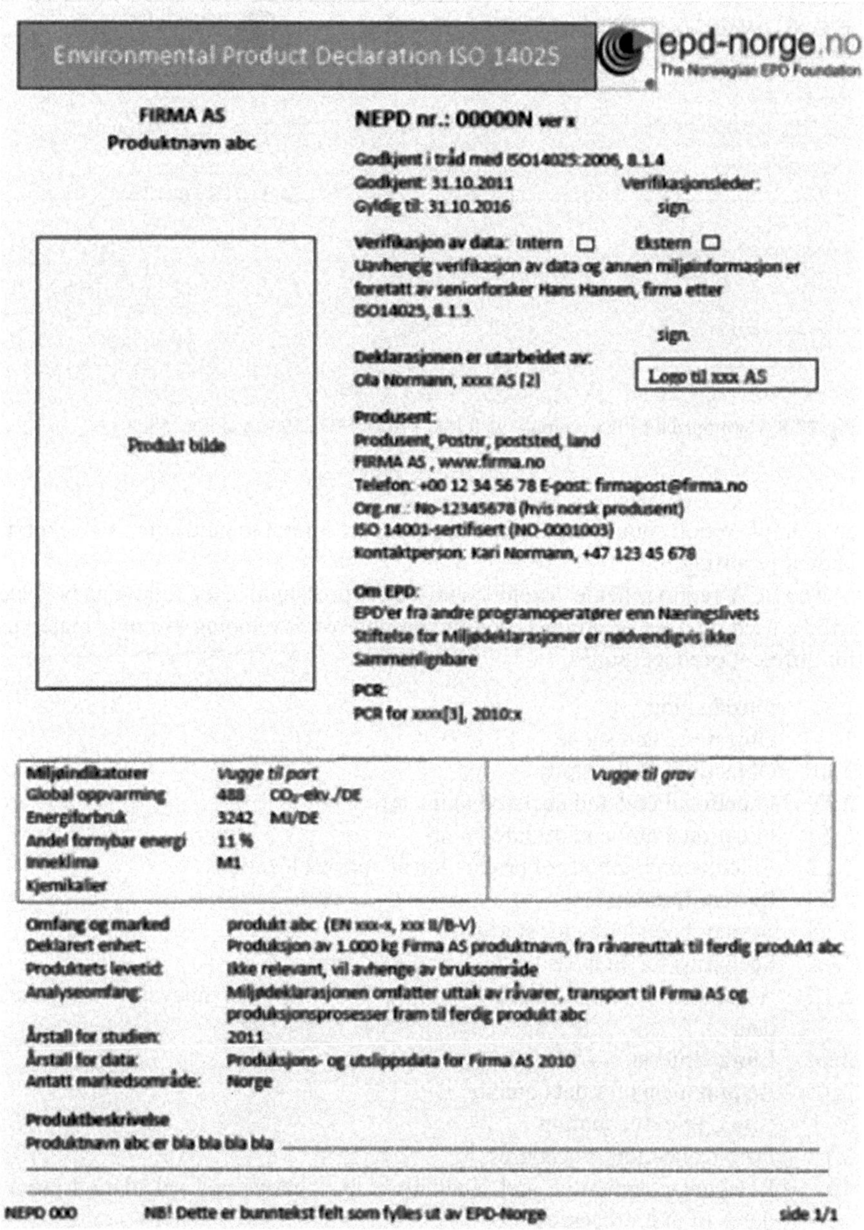

Environmental Product Declaration ISO 14025

epd-norge.no
The Norwegian EPD Foundation

FIRMA AS
Produktnavn abc

NEPD nr.: 00000N ver x

Godkjent i tråd med ISO14025:2006, 8.1.4
Godkjent: 31.10.2011 Verifikasjonsleder:
Gyldig til: 31.10.2016 sign.

Verifikasjon av data: Intern ☐ Ekstern ☐
Uavhengig verifikasjon av data og annen miljøinformasjon er foretatt av seniorforsker Hans Hansen, firma etter ISO14025, 8.1.3.
sign.

Deklarasjonen er utarbeidet av:
Ola Normann, xxxx AS [2]

Logo til xxx AS

Produkt bilde

Produsent:
Produsent, Postnr, poststed, land
FIRMA AS, www.firma.no
Telefon: +00 12 34 56 78 E-post: firmapost@firma.no
Org.nr.: NO-12345678 (hvis norsk produsent)
ISO 14001-sertifisert (NO-0001003)
Kontaktperson: Kari Normann, +47 123 45 678

Om EPD:
EPD'er fra andre programoperatører enn Næringslivets Stiftelse for Miljødeklarasjoner er nødvendigvis ikke Sammenlignbare

PCR:
PCR for xxxx[3], 2010:x

Miljøindikatorer	Vugge til port		Vugge til grav
Global oppvarming	488	CO_2-ekv./DE	
Energiforbruk	3242	MJ/DE	
Andel fornybar energi	11 %		
Inneklima	M1		
Kjemikalier			

Omfang og marked: produkt abc (EN xxx-x, xxx II/B-V)
Deklarert enhet: Produksjon av 1.000 kg Firma AS produktnavn, fra råvareuttak til ferdig produkt abc
Produktets levetid: Ikke relevant, vil avhenge av bruksområde
Analyseomfang: Miljødeklarasjonen omfatter uttak av råvarer, transport til Firma AS og produksjonsprosesser fram til ferdig produkt abc
Årstall for studien: 2011
Årstall for data: Produksjons- og utslippsdata for Firma AS 2010
Antatt markedsområde: Norge

Produktbeskrivelse
Produktnavn abc er bla bla bla bla

NEPD 000 NB! Dette er bunntekst felt som fylles ut av EPD-Norge side 1/1

Fig. 22.7 EPD front page

An important point of departure when this project started was that it should be a simple format while all requirements in the reference documents ISO 14044, point 5.2, ISO 14025, EN 15804, ands the ILCD handbook should be fulfilled. In addition

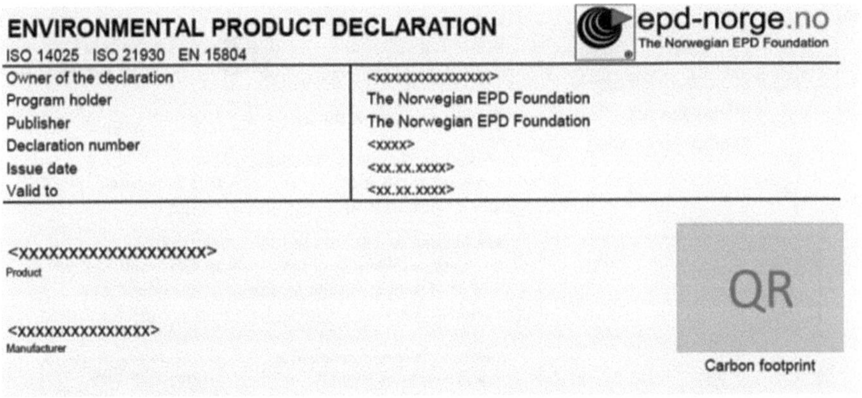

Fig. 22.8 Harmonized EPD to comply with ISO 14025, ISO 21930 and EN 15804

an example report should be attached to the report format to guide the LCA-experts in their reporting.

The LCA-report scheme developed in 2012 is presented in the following box and will be used on a set of product with the purpose of developing example materials for different product groups.

1. Introduction
2. Objectives and scope
2.1 Objectives of the study
2.2 Functional unit and declared unit
2.2.1 Functional unit and declared unit
2.2.2 Calculation methods of production at various locations
2.3 System Boundaries
2.3.1 System boundaries for study
2.3.2 Scenarios for analysis beyond cradle to gate
2.3.3 Assumptions about electricity production and other relevant background data
2.3.4 Limit Criteria
2.3.5 Requirements for data quality
3. Life Cycle Accounting
3.1 Data Collection Procedures
3.2 Relevant quantitative and qualitative descriptions and calculation procedures of unit processes
3.3 Sources of data used to conduct the analysis
3.4 Data Validation
3.4.1 Assessment of data quality
3.4.2 Handling of missing data
3.4.3 Quality Assurance Procedure
3.5 Allocation Policies and Procedures

3.5.1 Documentation and arguments for allocation procedures applied
3.5.2 Consistent application of the allocation procedures
4. Life Cycle Impact Rating
4.1 Impact assessment procedures, calculations and results from the analysis
4.2 Correlation between the results of impact assessment and inventory analysis
4.3 Characterization Models and Methods
4.4 Value Option
5. Life Cycle Interpretation
5.1 Interpretation of results
5.2 Assumptions and limitations of interpretation
5.3 Evaluation of data quality
5.4 Value-based choices and considerations
6. Critical Review
6.1 Participants' names and affiliations
6.2 Feedback from critical review
6.3 Reactions to recommendations
7. References
8. Appendix

22.6.4 The Format of EPD-Verification Report for the Norwegian EPD Foundation

In parallel to the project organized through the technical committee in The Norwegian EPD Foundation, the verification committee has developed a report form for the verification process. The Norwegian EPD Foundation has approved a group of experts as independent verifiers of EPDs, and this report form is to be used by the independent verifier to report to the verification committee that an EPD is verified and can be given a number in the Norwegian EPD-program. This will then allow the owner of this EPD to use the logo of EPD-Norway on their products.

In the verification report the verifier must report on the following topics: Name and information about the product the EPD is presenting, and Information about the producer. There are then two options; (1) if the report is a third party verification, or (2) if is a self-declaration of the verified results. According to the first option the verifier has to answer a series of questions about the underlying LCA and the quality of the data, and information about eventually databases that the products specific data is collected form. For the second option, the company itself can declare the quality of the LCA. However, this requires that the company is ISO 14001 certified and that they have procedures on how to perform LCA and to develop EPDs in their management system. They also need to state that if they do not have LCA-expertise within the company, they are collaborating with LCA-experts. If they do not have an ISO 14001 system, then they need to provide additional documentation for their LCA-expertise. This situation has not yet occurred in any company in Norway, meaning that uncertified companies quite seldom develop LCA and EPDs.

22.7 Summary and Conclusion

Since the establishment of The Norwegian EPD Foundation there has been an increasing demand for environmental information on products. This is among other based upon the requirements in Green Public Procurement. Green Public Procurement is a part of the national strategy for sustainable development and has become national policy through legislation and various initiatives from the government (Fet et al. 2011a). The legal basis is primarily three acts; the Public Procurement Act (1999), The Environmental Information Act (2003) and the Freedom of Information Act (2003).

The Public Procurement Act states that environmental considerations shall be taken in public procurement to ensure the most effective use of resources during public procurement and lead to the choice of products and services with low negative impacts on the environment. All governmental bodies must comply with this law when planning a purchase, including consideration of lifecycle costs, overall design and environmental impacts. This demand for environmentally friendly products and services should both encourage innovation and improve competitiveness in the markets that increasingly demand environmentally friendly goods and services.

The purpose of the Environmental Information Act (2003) is to ensure public access to environmental information, and people may demand information on life cycle environmental performance of any product. Especially the Public Procurement Act has been a driver for development of EPDs. This has further requested more information about how to prepare EPDs and also how to understand the information in an EPD. The guiding documents on PCRs, EPDs and LCA-reporting developed by The Norwegian EPD Foundation have been requested from different stakeholders and interested parties. The goal further is that they contribute to an increase in EPDs, but also to a more common understanding of the information provided in the declarations. As the regulations are requiring documentation on environmental aspects also for the common people, there is a need to make the information more available for the commons. The Norwegian EPD Foundation means that this is the way to go to achieve more sustainable products on the long term, and that the investments in the guiding documents is a step in this direction.

Another driver for further development of the EPDs is a national program on how to implement not only environmental responsibility but also Social Responsibility in Public Procurement. This means that the procurement in the public sector should be done with a minimum of environmental impact and with respect for fundamental human rights, and that goods and services are selected on the basis of life cycle costs, quality and the environment addressing energy efficiency, low content of hazardous chemicals, low pollution and low resource consumption. This stresses the focus on not only environmental but also ethical issues in public procurements. This addresses information about the suppliers' certificates, audits, compliance, reporting etc., and more extensive information about products through Eco-labels, LCA-reports, health aspects of products, content of chemicals or toxic material in products and end-of-life treatment of products. Some of this type of information is taken into the EPD-scheme that will be launched early in 2013, but still a lot remains for full information according to upcoming national and international requirements. The model that the Norwegian EPD Foundation has developed will be posted on the webpages with an open access.

References

Baumann H, Tillman A-M (2004) The Hitch Hiker's guide to LCA. Studentlitteratur, Lund

Bogeskär M, Carter A, Nevén C-A, Nuij R, Schmincke E, Stranddorf HK (2002) Evaluation of environmental product declaration schemes. European Commission DG Environment

CEN (2012) EN 15804: Sustainability of construction work—environmental product declarations—core rules for the product category of construction products. European Committee for Standardization, Brussels

Environmental Information Act (2003) Act relating to the right to environmental information and participation in decision-making processes relating to the environment, 9 May 2003, no. 31. Government Administration Services, Oslo

Fet AM, Skaar C (2006) Eco-labeling, product category rules and certification procedures based on ISO 14025 requirements. Int J Life Cycle Assess 11(1):49–54

Fet AM, Michelsen O, de Boer L (2011a) Green public procurement in practice—the case of Norway. Soc Econ 33(1):183–198

Fet AM, Rønning A, Folvik K (2011b) EPD generering med mål å tilstrebe sammenlignbarhet, EMCSR-report 8-2011. NTNU, Trondheim

Freedom of Information Act (2003) Act to public access to documents in the public administration, 20 June 2003, No. 45. Ministry of Justice and Public Security, Oslo, Norway

Hansen OJ, Stranddorf HK, Vold M, Lindfors LG, Hoffmann L (2001) Systems and methodology for type III environmental product declaration. Nordic Industrial Fund, Copenhagen

Heijungs R (2007) Analysis of functional chains: Life cycle assessment (LCA). In: Brattebø H, Røine K, Opoku H, Ehrenfeld JR (eds) Introduction to industrial ecology—theory, methods and applications. NTNU, Trondheim

Hillier B, Jonsson L, Ryding SO (2004) International guide to environmental product declarations. GEDnet, Stockholm

ISO (2004) ISO 14001: Environmental management systems—requirements with guidance for use. International Organization for Standardisation, Geneva

ISO (2006a) ISO 14025: Environmental labels and declarations type III Principles and procedures. International Organization for Standardisation, Geneva

ISO (2006b) ISO 14040: Environmental management life cycle assessment principles and framework. International Organization for Standardisation, Geneva

ISO (2006c) ISO 14044: Environmental management life cycle assessment requirements and guidelines. International Organization for Standardisation, Geneva

ISO (2007) ISO 21930: Sustainability in building construction—environmental declaration of building products. International Organization for Standardisation, Geneva

Kristensen P, Andersen L, Engelund B, Nielsen B (2006) Integrated supply chain information systems. Miljøministeriet, Copenhagen

Magerøy M (2011) The communication of environmental impacts through environmental product declarations, Master thesis. Norwegian University of Science and Technology, NTNU, Trondheim

Schenck R (2010) A roadmap to environmental product declarations in the United States. The Institute for Environmental Research and Education, IERE, Vashon

Chapter 23
Environmentally Sustainable Development in the Province of Belluno (Italy)

Veronica Novelli, Paola Geatti, Luciano Ceccon and Bruno Andrea Autelitano

Abstract In recent years, the municipality of Ponte nelle Alpi, in the province of Belluno, has developed an excellent environmental policy, that has allowed it to come first place in the 2010 Italian award "Comuni Ricicloni" for separate collections of household waste, to achieve the ISO 14001 certificate in 2010 and the Emas recognition in 2011. In 2010 the municipality started the joint project PubblichEnergie, together with other local authorities, to promote the use of renewable energy sources, in particular photovoltaic energy. Informative meetings with people, enterprises, associations and local authorities were convened on the opportunities of safeguarding both environment and public health. Among the innovative aspects of the project there is the possibility of supporting a Joint Purchase Group to develop the potentials of photovoltaic technology. Furthermore, the possibility of access to promotion initiatives granted by the EU, in particular the Energy Count, is provided for.

Keywords Ponte nelle Alpi municipality · Environmental policy · PubblichEnergie project · Photovoltaic energy · Joint purchase groups

23.1 Introduction

Demographic growth and related industrial and commercial activities result in negative effects on living conditions on our planet. In particular, the use of fossil fuels increases environmental pollution; therefore, it is fundamental to use renewable energies and related conversion technologies. Present reserves of oil and natural gas can only cover consumption at this rate for respectively the next 50 years in the case of oil and 70 years in the case of natural gas. The quality of life and security

V. Novelli (✉) · L. Ceccon
Department of Economics and Statistics, University of Udine, 33100, Udine, Italy
e-mail: novelli@uniud.it

P. Geatti
Department of Chemistry, Physics and Environment, University of Udine, 33100, Udine, Italy

B. A. Autelitano
University of Udine, 33100, Udine, Italy

R. Salomone, G. Saija (eds.), *Pathways to Environmental Sustainability*,
DOI 10.1007/978-3-319-03826-1_23, © Springer International Publishing Switzerland 2014

of present and future generations depend strongly on the use of renewable energies (Stambouli et al. 2012). The reduction of anthropogenic impacts on the environment is a priority. The concept of sustainability has always been important for some nations: a wise North America Indian motto recites: "We did not inherit the Earth from our ancestry; we borrowed the Earth from our children" (Milutiene et al. 2012). Sustainability will be the main focus for the field of public administration in the coming decade (Fiorino 2010); many municipalities already pay great attention to the environment (Sharifi and Murayama 2013; Daddi et al. 2011; Fiaschi et al. 2012; Mainero 2012; Patlitzianas and Christos 2011; Ǿstergaard and Lund 2011; Ǿstergaard et al. 2010; Federico et al. 2009).

Ponte nelle Alpi (PnA) is a small town in the province of Belluno nestling in the mountain/hilly countryside; it is connected by rail to Belluno, Padua and Venice and by the A27 motorway to Venice. To pursue its environmental objectives, PnA has decided to participate in various initiatives. The main goal is sustainable territorial and commercial development for its inhabitants and future generations. Significant actions are the implementation of environmental policies, such as environmental quality certification, better management of water, soil and subsoil, air, noise, urban wastes, urban lighting, a better use of renewable energies (Autelitano 2012).

The aim of this chapter is to evaluate the environmental performances of PnA, considering in particular the use of photovoltaic energy.

23.2 The Environmental Policy

The municipality of PnA includes in its system of environmental management the organizational structure, activities, responsibilities, procedures and processes for a policy based on an ongoing, aware improvement of environmental performance. PnA reduces pollution by controlling the environmental impact of various activities, lowering production and stimulating separate collections of household waste, favoring sustainable tourism and use of renewable energies, mainly photovoltaic (PV) energy.

23.2.1 Air Quality

From 2001 to 2006 Arpav (Agenzia Regionale per la Prevenzione e Protezione Ambientale del Veneto; Environmental Protection Agency of the Veneto Region) controlled the air quality of PnA by using a mobile laboratory and measuring the concentrations in the atmosphere of pollutants such as sulfur dioxide, nitrogen dioxide, benzene, toluene and xylene; the values were below the threshold limits imposed by regulations. In more recent years, values of PM 10, PM 2.5 and carbon monoxide have also been checked; the monitoring of ozone is provided for, as well.

23.2.2 *The Purification of Wastewater*

PnA controls and verifies the correct working of the service of purification of wastewater by means of regular physical, chemical and microbiological analyses.

23.2.3 *Noise*

PnA grants Anas (Agenzia Nazionale Autonoma delle Strade Statali; National Agency of State Roads) and the province of Belluno control of the level of acoustic pollution due to vehicular traffic, in order to respect the current regulations, while Arpav is engaged in noise measurements when excess is expected.

23.2.4 *Urban Lighting*

As far as energy saving and consequent expense reduction are concerned, PnA has scheduled a gradual passage to lighting with low dispersion of light. Such a pathway has been made possible by implementing a new system and adapting existing ones. These changes have been carried out on the roads and in green areas.

23.2.5 *Household Waste Management*

In 2006 the separate collection of household waste was 22.4% and per capita production was 348 kg/year. Costs related to disposal of undifferentiated waste were const on the increase. Waste production had to be reduced to guarantee a lower cost for disposal. Another goal consisted in avoiding the building of incinerator systems. By controlling the costs of services delivered, it was possible to allow citizens to pay rates on the waste produced. In May 2007 PnA decided to activate door-to-door service of separate collection of household waste, providing for the subdivision into dry material, wet products, glass-plastic-aluminium. Citizens were informed of these changes and PnA instructed its inhabitants on their new tasks. The municipal Ecocentro was inaugurated: it was the first authorized centre open to industrial plants, too, and predisposed for thirty different types of waste products.

23.2.6 *Management of the Municipal Patrimony*

The maintenance office of PnA ensures that the fire fighting services of its buildings and structures (the Town Hall, schools, public library, theatre, old people's home, municipal garage, indoor stadium, football field) are efficient and maintained regularly according to law. Analysis of emissions and the output of heating systems is also carried out regularly by an external company, the activity of which is verified by Reports and Minutes.

23.2.7 Territory Planning

This activity includes the editing of Town Plans. At present the guidelines for the territorial development (Piano di Assetto del Territorio, P.A.T.) for the next ten years are being put in place. In 2008 PnA and the Veneto Region signed an agreement of co-planning, extended in 2009 to the Province of Belluno. Full autonomy is given to each administration about active participation by interested subjects, such as city dwellers, hamlet leaders, cultural associations, sport and local associations. An important phase of P.A.T. is the programming of the Intervention Plan, a fundamental document for the General Town Plan.

23.2.8 Participation and Environmental Communication

PnA's environmental policy provides for informative meetings for municipal employees to grasp the importance of environmental topics. The mayor, the municipal council and the councillors are involved. The implementation of the Environmental Management System (E.M.S.) consists of many activities; roles of responsibility inside the municipal administration, as the Environmental Delegate, the Operative Referent of the E.M.S. and the Person Responsible for the Supervision and Environment, have been set out. Several initiatives have been activated by PnA in terms of transparency, to involve and to inform the inhabitants about the excellent results obtained; meetings related to different themes, such as wastes, renewable energies and bio-diversity have also been arranged. As far as waste is concerned, lessons of environmental education are promoted in the schools, to develop a virtuous education in young people. Instructional guided tours of Ecocentro, of Centro Riciclo Vedelago in the province of Treviso and didactic laboratories are scheduled. As far as bio-diversity is concerned, PnA has activated initiatives, suggested by the inhabitants, such as "Zanzare e pipistrelli" ("Mosquitos and bats"), that recognizes the ability of the latter to reduce pesticide use, and the "Giornata dell'albero" ("Tree day"), when in some parts of the municipal area several trees are planted to preserve the equilibrium of the ecosystem.

23.3 Other Initiatives, Subscriptions, Recognitions and Awards

PnA has built a centralized plant for the distribution of liquefied oil gas and has replaced old diesel oil boilers with condensing boilers, alimented by natural gas. Taking into account the frequent problems of water deficiency, rain water harvesting and re-use plants have been built: PnA has also published an announcement of grants available for farmers to provide for such problems. In 2011, PnA took part in the municipality network "Alleanza nelle Alpi" ("Alliance in the Alps"), which, together with seven other municipalities, aims at sustainable development in the alpine area, by sharing information, knowledge and advice and taking part in local, national and international events. Moreover, PnA joined the project "Su e so par i coi" (Up and down the hills), with the purpose of meeting the needs of old people having problems of mobility, allowing them to reach both main market places and public services. PnA also intends to use at least 70 % of recycled paper in its offices, instead of the 30 % provided for by current Italian rules (Ministry of Environment 2003), and 30 % of other "green" products. PnA attention towards environmental themes is proved by the ISO 14001:2004 certification and the renewal of the EMAS registration on 4th October 2011. Compatibly with the ISO 14001:2004 certification, PnA obtained the international certification IQNet for implementation and maintenance of its E.M.S. PnA has also obtained various nationwide recognitions and awards.

23.3.1 "Comuni a 5 stelle" ("Five-star municipalities") Award

PnA was the outright winner of the national 2008 award "Comuni a 5 stelle", promoted by the "Associazione Nazionale Comuni Virtuosi" (Virtuous Municipalities National Association). The main aim was to identify a tool to stimulate and render responsible the administrations in order to reduce district ecological footprint. The award was also an opportunity to foster the birth of citizen ecological awareness and afterwards to improve environmental conditions (Associazione dei Comuni Virtuosi 2012),

23.3.2 "Comuni Ricicloni 2010" ("The Most Efficient Waste-Recycling Municipalities in 2010") Award

PnA obtained the prestigious recognition "Comuni Ricicloni 2010" gaining first place in Italy for separate collections of household waste.

23.3.3 Energimed 2009 Award

In March 2009, PnA obtained the "Energimed 2009" third edition award (coming first *ex aequo* to the Bellizzi municipality, in the province of Salerno) for appropriate activities of municipal administration on renewable energies and waste recycling.

23.3.4 *Life AgEMAS Project Award*

In June 2009, the E. C. awarded PnA, together with the Dolomiti Bellunesi National Park and the Belluno, Feltre and Pedavena municipalities and the Agordina valley, for the Life AgEMAS project (Comune di Ponte nelle Alpi 2012).

23.4 Use of Renewable Energy Sources

Among the various renewable energy sources, PnA dedicated particular attention to PV installations. The technology allows the transformation of solar energy directly and immediately into electrical energy without the utilization of fossil fuels (ENEA 2006). Solar energy is the most important energy source for life on the planet and also the cleanest from the environmental point of view. Its use prevents carbon dioxide and other pollutants, damaging to the environment and to health, to be released into the atmosphere. PV systems are basically classified in isolated systems (stand alone) and grid-connected systems. In the former case, the energy produced exceeding consumer needs is cumulated in specific batteries. PV plant is not grid-connected and can be assembled with electric generators or wind-generators. On the other hand, grid-connected systems pour yielded energy into the electricity grid they are connected to. Possible advancements of PV systems rely on research on new materials that could be used in place of silicon (Renael 2009). There are various public incentives to use PV energy. These include green certificates, negotiable annual bonds which certify the energy production from renewable sources; collected energy, that provides for the market placement of the produced energy through the brokering of the *Gestore dei Servizi Elettrici* (Electrical Service Manager); on-the-spot exchange, which measures the value of the energy input of the electrical grid on the basis of an economic balance criterion with the value of the energy output from the electrical grid; the Energy Count, an inducement mechanism for energy production from solar source (Rubini and Sangiorgio 2012). The first project realized by PnA, concerning the production of electrical energy by PV technology, was started in 2006 and finished in 2008, when, in the junior high school, a 25 kWp plant was installed. As a consequence of the success obtained from both the environmental and economic point of view, PnA decided to plan and realize further PV plants on municipal structures having suitable roofs or coverings. In subsequent years several plants were installed in various areas of the municipal district, for example on the nursery school at Piaia, on the primary school at Ponte nelle Alpi, on the sports arena, on the multipurpose area platform at Polpet, on the junior high school at Canevoi and over the graveyard at Col di Cugnan. As a consequence, a 218 kW production of PV energy is estimated on a surface of approximately 1,900 m^2, when the connection is complete. Moreover, the installation of a plant near the municipal stadium is being planned (Consorzio Bim Piave 2012).

23.5 The PubblichEnergie Project

PubblichEnergie is a project which started in 2010 on the initiative of the municipalities of Ponte nelle Alpi, Alano di Piave, Mel, Lentiai and Trichina, in the province of Belluno, and under the aegis of the Mountain Consortium of Belluno – Ponte nelle Alpi. The project, centered on PV technology, is based on energy saving and sustainable mobility and offers important chances to both public and private parties of the territory. Both the municipality of Agordo and the Agordina and Pedavena Valley joined the initiative in 2011; the municipality of Valle di Cadore joined in 2012 (Comune di Agordo 2012).

A convention is signed every year by all the governments adhering to the convention as to the purposes, the duration, the economic relationships, mutual duties and guarantees. This document represents the testimonial of the service supply to the people with reference to promotion of both renewable energies and energy saving.

PnA is the leader of the initiative and is delegated to indicate:

- the actor to carry out the plan, i.e. the "PubblichEnergie" common Office;
- the regulations relative to the activities and all the organizing aspects to be submitted for the approval of the town councils.

23.5.1 The PubblichEnergie Office

The PubblichEnergie Office avails itself of the collaboration of the staff of the offices of the annexed municipalities, to guarantee the supplying of data and information necessary for the service. In case of need, it can take advantage also of the contribution of external technical consultants. The PubblichEnergie Office has the main purpose of "developing the potentialities of sustainable economy at the local level, for saving and consumption quality for families, growth for enterprises, service for local authorities and their residents, defence for environment and public health" (PubblichEnergie 2012).

The main activities carried on by the Office are:

- information and advice to citizens, enterprises, fellowships, associated and non-associated local authorities;
- promotion and organization of procedures to increase the purchasing power of families and enterprises;
- organization of activities by continuously informing the associated municipalities and co-ordinating with them for the initiatives addressed to the respective citizens;
- promotion of interventions for energy saving and the production of renewable energies in favor of families, enterprises and real estates of the associated authorities;

- setting up of development opportunities for enterprises and training opportunities for students, workers and technicians;
- improvement of the local environment.

Collaboration is necessary on the part of associated municipalities to achieve expected results and to supply the necessary support to the common Office for carrying out the activities addressed to the citizens.

23.5.2 Financial Relations

A cost sharing initiative is provided for the associated municipalities; costs can be lowered in proportion to the receipts that the common Office is able to obtain for the activities that are implemented addressed to citizens, enterprises, fellowships and other local authorities. Budget management is given to PnA, which has to compile the final statement of accounts. If the common Office achieves a credit balance, this will be distributed between the associated municipalities in proportion to the number of residents, thus lowering the cost contribution of each municipality to its zeroing. Possible further resources collected by the common Office will be assigned to activities and services offered to the citizens of the municipalities taking part in the project. If outcomes of financial management do not break even, the amount of debit will be distributed among the associated municipalities in proportion to the number of residents, thus increasing the cost contribution of each municipality.

As a rule, the associated municipalities will give the final amounts due to PnA, pursuant to the statement of accounts produced by the leader municipality. The possibility that other municipalities may agree to the project during its term with the status of "partner municipalities" is provided for; the partner municipalities will benefit from the same services offered to the other municipalities. However, the participation in the division of possible profits is not provided for the partner municipalities. There is the possibility for a municipality to back out of the agreement giving at least 90 days notice.

23.5.3 Further External Supports

Since 2011 PubblichEnergie benefits from the support of "DYNALP di Alleanza nelle Alpi", a supranational consortium of Alps municipalities that provides for the presentation of the project to other frameworks, also international, and a profitable exchange of experiences to enrich the offer (http://www.alpenallinaz.org/it/progetti). From 2012 other Italian municipalities may agree to the initiative, as well, in collaboration with the "Arianova Sostenibile" onlus cooperative. Thus the PubblichEnergie Italian network was born and now comprises 9 municipalities of the province of Parma (Busseto, Collecchio, Colorno, Felino, Mezzani, Montechiarugolo, San Secondo Parmense, Sorvolo and Traversetolo). The initiative is sponsored by the "Associazione Nazionale Comuni Virtuosi" (Virtuous Municipalities National Association), and its promoters plan to develop the model at the national level. Furthermore, PubblichEnergie takes part in the "Scuola di AltrAmministrazione", promoted by the "Associazione Nazionale Comuni Virtuosi" favoring the exchange of good practices and initiatives among the associated municipalities (Scuola di Altra Amministrazione 2012).

23.5.4 The Innovative Aspects of the Project

One of the main aspects of the project is the possibility of supporting Joint Purchase Groups (JPGs), that are widespread in the voluntary world, as public initiatives to develop the derived potential. This opportunity provides for the satisfaction of the common interest by both guaranteeing transparency of action and favoring the participation in the decision-making processes by citizens and developing actions that allow knowledge circulation and effective application of European, national and regional rules for citizen and territory protection. The main innovative aspects of the project are:

- the organizing model is simple and cheap, but effective, as it aims at favoring encounters with citizens and their participation. Therefore, the setting up of a service addressed to the community, resolving problems, favoring and urging citizens to join initiatives, the opportunities that may be exploited, the possible benefits, not confined to "waiting for citizens" is provided for. An active model contrasting the typical approach by the offices is thus developing;
- the JPG methodology ensures the respect of principles of transparency and impartiality. The main purpose is to favor the making of the PV plants for the participating citizens, by lowering building times and costs and by offering quality and service guarantees that are on average higher than those usually available on the market;
- the citizens are informed of the possibility of exploiting current economic incentives and are facilitated in optimizing their investment decisions. People have often a limited, and in some cases inadequate knowledge of the advantages that derive from State incentives, laws and decrees. This is partly due to a widespread lack of interest by the people and partly due to inefficacious presentation and advertising by the State. This project aims at reducing such a gap and increasing knowledge and information of issues that might be favorable from both an environmental and economic point of view. With its activities and informative meetings, PubblichEnergie aims at pressing people to become active parties of the system by favoring their participation;
- the local economy, and particularly the fields of green economy or "responsible economy", showing great growth potentials, are stimulated. The project helps the development of local associated activities by stimulating the use of the PV technology;

- the collaboration between municipalities is aided by the establishment of the common Office for the carrying out of non essential, but value-added service;
- the full covering of the costs of service is guaranteed, as supporters of the JPG transfer a small part of the savings obtained by shared purchase;
- communication is based on the starting up of informal social networks that encourage economical dissemination and advertising, e.g., by word of mouth. Another important aspect is relative to the initiatives promoted to favor media interest towards the project;
- mutual understanding on the part of the municipalities involved is fundamental, so that all actors participating in the project share ideas, ways and initiatives that are supported by ongoing co-ordination at councillor level and provide for widespread growth.

These eight points underpin a project ongoing at the local level, in the prospect of diffusion at the national level (Orzes 2011).

23.5.5 Expected Targets

The two main targets to which the project is addressed comprise:

- citizens, who obtain value-added information and to whom the advantages are illustrated of investing in renewable energy resources, in energy saving interventions for houses and in sustainable building. The main purpose is to help families to make the best decisions and to support them by the voluntary participation in a JPG allowing them to reduce purchase costs and to obtain better contract guarantees;
- territory, by helping local enterprises to offer such improvements, from both an economic and qualitative point of view. Combined activities, the Alps area development and job creation are favored. Furthermore, pollutant release into the environment is reduced by even minor structural interventions in house building and other buildings (Orzes 2011).

23.5.6 The Development Steps of the Project

23.5.6.1 Preliminary Step

The first step was the start of the contacts between the municipalities interested in the project by drawing up and arranging preliminary meetings. An agreement draft for the setting up of a common Office was then prepared. The budget and the organizing details were defined, including the schedules of the informative and advertising meetings, and the times and ways of opening of a "citizen window". After the associated municipalities passed the agreement, the drawing up of the final copy was provided for and the leader municipality identified.

23.5.6.2 Promotion

The next step provided for logo drafting, website implementation, planning and carrying out of informative material such as brochures, playbills and roll-ups. Furthermore, sponsors were looked for and press conferences were organized to disseminate the project in the presence of journalists and by means of press releases.

23.5.6.3 Start of the Activities

Weekly informative thematic evenings were carried out in the both associated and non associated municipalities. These initiatives were open to everyone in the territory and enterprises and aimed at explaining the purpose of the project and PV technology. The Energy Count bonus, the JPG concept and its organizing steps were also explained. A presentation was prepared and illustrated to facilitate understanding. The opening of the "citizen window" in all the associated municipalities was achieved, where weekly meetings were scheduled for further advice and explanations. The "citizen window" fulfils three purposes:

- to replicate the evening presentations to small groups of residents, often housewives or workers who can't participate during evening time;
- to estimate specific intervention scenarios for people joining a PV JPG;
- to represent a point of reference for private citizens and public authorities.

23.5.6.4 Organization and Management of a JPG

The following steps are involved:

- collection of preliminary adhesions and training of the group members by informative meetings;
- process of "informal bidding" with public tenders in conformity with the methods and choice of suppliers by the JPG supporters;
- formal award of a "general agreement" between the JPG and the supplying companies;
- drawing up of the typical agreement between each participant citizen and companies, on the basis of the general agreement;
- start of technical on-the-spot inspections, agreement drawing up and plant installation. Ongoing monitoring of the respect of settled conditions, of critical situations and of associated citizen satisfaction is provided for;
- realization of the "informal bidding" with the banks of the territory to draw up an agreement by which associating citizens can obtain favorable conditions of financial support;
- carrying out of market research to identify the best policy satisfying the needs of the JPG supporters to insure their PV plants;
- contribution collection by supporters and enterprises.

23.5.6.5 External Communications

They provide for ongoing dialogue, open comparison and active communication with category associations of enterprises, with territory institutions, with local mass media and the participation in conventions, exhibitions and debates. Furthermore, a basic requirement is the constant updating of the website, so that the project characteristics can be disseminated.

23.5.6.6 Co-ordination Among Administrations

A Councillor of the leader municipality is the coordinator of the project and calls regular, usually two-monthly, meetings. Ongoing monitoring of the project is provided for and guidelines on general matters and external communications are settled by participation of the Mayors and/or of Councillors of the associated municipalities.

23.5.7 Results Reached by PubblichEnergie

By mid 2012, the project had arranged 35 informative evenings and 28 months of citizen window to its credit. The citizens contacted numbered about 3,500; the plants installed on private houses numbered 270, and other plants had been installed on small enterprises, altogether in 26 municipalities of the provinces of Belluno and Parma. The total power installed was about 1,100 kWp and the yearly energy produced about 1,200,000 kW/h. Six hundred tonnes per year of atmospheric emission of carbon dioxide are estimated to have been avoided. Twenty-two subjects including enterprises, professionals and craftsmen were involved in the initiative. Total finances involved amounted to about € 3.8 million.

23.5.8 Further Activities and Initiatives

PubblichEnergie has started further activities and initiatives. The Energianova project related to house energy saving, is addressed to citizens and is aimed at offering basic knowledge for understanding critical points responsible for main house heating costs. People may then contact their own technicians to find possible solutions. The initiative is supported by B.I.M. PIAVE Co-operative and consists of three levels:

- small training courses, with free admittance, in the municipalities participating in the project;
- detailed analysis of individual situations, by means of questionnaires and carrying out technical on-the-spot inspections;
- support in the operative decisions, also by possible establishment of JPGs.

Furthermore, PubblichEnergie promotes a sustainable mobility project providing incentives for cycle use for short journeys and for cycle-path infrastructure. Finally, PubblichEnergie organizes activities and meetings, handing out a guide-book for customers and enterprises, to spread and help safety in job places during installation of the PV plants (PubblichEnergie 2012).

References

Associazione dei Comuni Virtuosi (2012) Ponte nelle Alpi vince il premio "Comuni a 5 stelle". http://www.comunivirtuosi.org/edizione-2008/premio-comuni-a-5-stelle/edizione-2008/ponte-nelle-alpi-vince-il-premio-comuni-a-5-stelle. Accessed 20 Nov 2012

Autelitano BA (2012) Ponte nelle Alpi e lo sviluppo eco-sostenibile del suo territorio. Dissertation, University of Udine (Italy)

Comune di Agordo (2012) PubblichEnergie. http://myportal.regione.veneto.it/opencms/comune.act?news=/opencms/opencms/CMA/Agordo/PrimoPiano/PubblichEnergie.html. Accessed 20 Nov 2012

Comune di Ponte nelle Alpi (2012) http://www.comune.pontenellealpi.bl.it/xhtml/ufficioambiente. Accessed 20 Nov 2012

Consorzio Bim Piave (2012) www.consorziobimpiave.bl.it/web/bim. Accessed 20 Nov 2012

Daddi T, Frey M, Iraldo F, Nabil B (2011) The implementation of an environmental management system in a North-African local public administration: the case of the City Council of Marrakech (Morocco). J Environ Plan Manag 54(6):813–832

ENEA (2006) L'energia fotovoltaica. Collana Sviluppo Sostenibile. http://old.enea.it/produzione_scientifica/pdf_op_svil_sost/Op22.pdf. Accessed 20 Nov 2012

Federico G, Rizzo G, Traverso M (2009) In itinere strategic environmental assessment of an integrated provincial waste system. Waste Manag Res 27(4):390–398

Fiaschi D, Bandinelli R, Conti S (2012) A case study for energy issues of public buildings and utilities in a small municipality: investigation of possible improvements and integration with renewables. Appl Energy 97:101–114

Fiorino DJ (2010) Sustainability as a conceptual focus for public administration. Public Adm Rev 70(SI):S78–S88

Mainero D (2012) Integrated anaerobic-aerobic systems for solid waste treatment. Rev Environ Sci Biotechnol 11(4):343–352

Milutiene E, Staniskis JK, Krucius A, Auguliene V, Ardickas D (2012) Increase in buildings sustainability by using renewable materials and energy. Clean Technol Environ Policy 14:1075–1084

Ministry of Environment (2003) Ministerial Decree 08/05/2003 no. 203—Norme affinché gli uffici pubblici e le società a prevalente capitale pubblico coprano il fabbisogno annuale di manufatti e beni con una quota di prodotti ottenuti da materiale riciclato nella misura non inferiore al 30 % del fabbisogno medesimo. Official Gazette of the Republic of Italy no. 180, 5 August 2003

Orzes E (2011) PubblichEnergie. http://www.altramministrazione.it/altra2assets/uploads/Ponte-nelle-Alpi.pdf. Accessed 20 Nov 2012

Østergaard PA, Lund H (2011) A renewable energy system in Frederikshavn using low-temperature geothermal energy for district heating. Appl Energy 88:479–487

Østergaard PA, Mathiesen BV, Moller B, Lund H (2010) A renewable energy scenario for Aalborg Municipality based on low-temperature geothermal heat, wind power and biomass. Energy 35:4892–4901

Patlitzianas KD, Christos K (2011) Sustainable energy investments in hellenic urban areas: examining modern financial mechanisms. Renew Sustain Rev 15:5186–5193

PubblichEnergie (2012) http://www.progettobenesserecompleto.it/serate-per-gas-per-fotovoltaico-e-risparmio-energetico/#.UUG9AI6wt5k. Accessed 20 Nov 2012
Renael (2009) Vademecum fonti rinnovabili. Energia solare. http://www.renael.net/public/documenti/181/SOLARE%20RENAEL%202007.pdf. Accessed 20 Nov 2012
Rubini L, Sangiorgio S (2012) Le energie rinnovabili. Hoepli, Milano
Scuola di Altra Amministrazione (2012) La Scuola di AltRa Amministrazione. http://www.altramministrazione.it/la-scuola-di-altra-amministrazione/. Accessed 20 Nov 2012
Sharifi A, Murayama A (2013) A critical review of seven selected neighborhood sustainability assessment tools. Environ Impact Assess Rev 38:73–87
Stambouli AB, Khiat Z, Flazi S, Kitamura Y (2012) A review on the renewable energy development in Algeria: current perspective energy scenario and sustainability issues. Renew Sustain Energy Rev 16:4445–4460

Chapter 24
The European Ecolabel in the Tourist Sector: An Analysis of the Italian Experience of Mountain Huts

Barbara Campisi, Federica Marinatto and Paolo Bogoni

Abstract In tourism and hospitality service industry numerous are the voluntary eco-labels that have been developed by individual companies, industry associations, voluntary organizations and government agencies. Their scopes can range from single villages to worldwide or from single activities to entire destinations; and they may include voluntary codes, awards, accreditation and certification schemes. However, the eco-labels are a controversial topic in tourism as the degree to which they affect tourists' planning processes and corporate environmental performance is largely unknown. This chapter mainly addresses the perception issue about the main difficulties and advantages driven by the adoption of the voluntary scheme based on the EU Ecolabel Regulation for the environmental certification of the product group 'tourist accommodation service'. In particular, here are presented the first results of a preliminary study carried out in a specific context: the mountain huts.

Keywords Sustainable development · Voluntary environmental initiatives · European Ecolabel · Tourism accommodation service · Mountain huts

24.1 Introduction

Since the early 1990s in tourism and hospitality service industry several have been the self-regulation and voluntary initiatives developed in support of sustainable tourism to minimize its environmental and socio-cultural impacts. Their scopes can range from single villages to worldwide, or from single activities to entire destinations; and they may include voluntary codes of conduct, awards, accreditation, and certification schemes. In particular, in many of the certification schemes governments play an important role given the opportunities that these initiatives

B. Campisi (✉) · P. Bogoni
"Bruno de Finetti" Dept. of Economics, Business, Mathematics and Statistics,
University of Trieste, 34127 Trieste, Italy
e-mail: barbara.campisi@econ.units.it

F. Marinatto
University of Trieste, 34127 Trieste, Italy

R. Salomone, G. Saija (eds.), *Pathways to Environmental Sustainability*,
DOI 10.1007/978-3-319-03826-1_24,

may provide in attaining sustainable development policy goals. The logos, included in certification and ecolabelling schemes with strict compliance and performance criteria, have the potential to influence production and consumption patterns along the whole tourism supply chain. In this way, the host society, the environment, governments, private companies and consumers may benefit from their operating as studies on the existing voluntary initiatives has pointed out their high potential to move the tourism industry towards sustainability (UNWTO 2002). For that reason governments are recommended to support in various ways the establishment and/or the application of national certification systems for sustainable tourism development (UNWTO 2003).

However, the eco-labels are a controversial topic in tourism as the degree to which they affect tourists' planning processes and corporate environmental performance is largely unknown (Buckley 2002; Chen and Peng 2012). On the contrary, it is common thinking that the proliferation of ecolabelling schemes, also in tourism, has undoubtedly generated confusion among consumers and without adequate marketing and education, the effectiveness of these programs may be rather limited. Many studies have demonstrated that generally there is a poor public awareness and knowledge about existing eco-labels and award schemes and hence an incorrect or incomplete understanding of their meaning, particularly among the customers.

Amongst the voluntary initiatives involving the development of environmental labeling standards, there are the third party ecolabelling schemes, based on life cycle considerations, introduced by national standards, such as the German Blue Angel and the Nordic Swan, or by the European Regulation issued for the first time in 1992 and lastly revised in 2009 (European Parliament and the Council of European Community 2010). The EU Ecolabel scheme is part of the sustainable consumption and production policy of the European Community, linked with other instruments, such as Green Public Procurement (GPP), the Eco-Management and Audit Scheme (EMAS), the Ecodesign directive, etc. In particular, it is aimed at reducing the negative impact of consumption and production on the environment, health, climate and natural resources through the promotion of the use of the EU Ecolabel logo (the Flower) for environmentally sound and high performance products. Currently, the European scheme covers 26 product groups, including tourist accommodation and camp services. The criteria for services were initially approved in July 2001 after that the European Commission had commissioned ANPA (Italian Environment Protection Agency, now Institute for Environmental Protection and Research or ISPRA) to develop a draft criteria proposal establishing the ecological criteria for the award of the EU Ecolabel to 'Tourist Accommodation Service' in collaboration with the Greek Competent Body, ASAOS. The criteria were adopted with Commission Decision of 14 April 2003 (2003/287/EC). In 2003 also the development of ecological criteria for 'Camp Site Service' were commissioned by the European Commission to APAT (Italian National Agency for the Protection of the Environment and for Technical Services—now ISPRA) which were adopted with Commission Decision of 14 April 2005 (2005/338/EC) (ISPRA 2009a). In 2009 the new revised European mandatory and optional ecological criteria for tourist accommodation and camping site services were adopted (2009/578/EC and 2009/564/EC—Commission

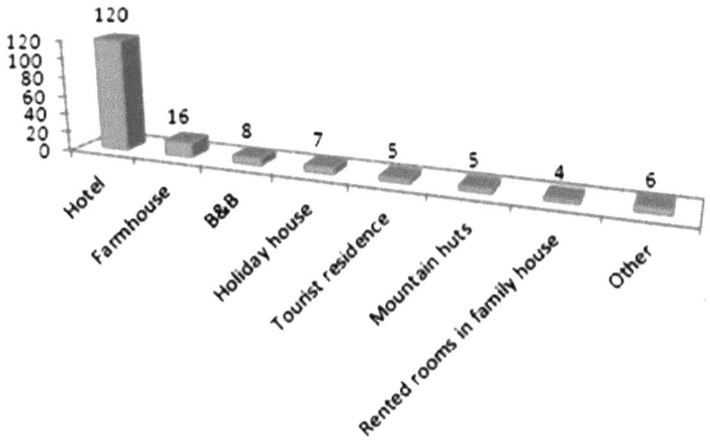

Fig. 24.1 Number of Italian EU Ecolabel awards for tourist accommodation service in September 2012. (Source: ISPRA 2012)

Decision of 9 July 2009) (Commission of the European Communities 2009). In the same year ISPRA also issued a User Manual to assist the Tourist accommodation service providers during the application process (ISPRA 2009b).

The accommodation products eligible for the EU Ecolabel may vary within a range of services. In Fig. 24.1 the number of EU Ecolabel awards (licenses) for tourist accommodation service in Italy, as of September 2012, is represented for service typology: hotels, farmhouses, bed & breakfast (B&B), holiday houses, tourist residences, mountain huts, rented rooms in family house, etc. "Other" includes the following typologies: apartment, 'albergo diffuso', guest house, country house, holiday camp, and hostel (ISPRA 2012).

Undoubtedly, according to the European statistics the inclusion of the tourist accommodation service among the product group has increased the demand for use of EU Ecolabel both in Italy and in the rest of Europe (Eurostat 2011). Between 2000 and 2010 the total number of Ecolabel licenses increased by 36.1 % per year on average, i.e. a multiplication by a factor of 21.8 over this period; and this results was due particularly to the tourism accommodation service sector where the number of licenses are still increasing rapidly. However, it should be noted that this increase does not correspond in an augmentation of market share of the related products. It is for that reason that in order to favor a further development of the scheme the EU Ecolabel Work Plan for 2011–2012 was defined with the aim at further increasing the interest for the European Flower (The Commission of the European Communities 2011). The definition of a precise strategy for the development of the scheme implies also the definition of targets and concrete actions to monitor and manage this process.

This chapter mainly addresses the perception issue about the advantages driven by the award of the Community eco-label to tourist accommodation service and the main difficulties encountered by the applicants. In particular, here are presented the

Table 24.1 EU ecolabelled mountain huts in Italy. (Source: ISPRA Ecolabel EU Servizi Certificati - May 2012)

Name of the structure	Site	Altitude (m)	Region
Arlaud	Montagne Seu (TO)	1,770	Piemonte
Toesca	Pian del Ròc (TO)	1,710	Piemonte
Federici-Marchesini al Pagarì	Entraque (CN)	1,587	Piemonte
Pian delle Gorre	Chiusa Pesio (CN)	1,032	Piemonte
Bertagnoli	La Piatta (VI)	1,225	Veneto
Flaiban-Pacherini	Forni di Sopra (UD)	1,587	Friuli Venezia Giulia

preliminary results of a study aimed at evaluating the experience of the participation in the EU Ecolabel scheme in a specific context of tourist infrastructure: the mountain huts.

24.2 Methodology

This qualitative study was performed by using a semi-structured questionnaire involving 22 closed questions and 6 open-ended questions in order to collect verbatim comments about the specific experience, benefits and difficulties related to the EU Ecolabel, along with practices adopted for the environmental management and external communication. The designed questionnaire was aimed to gather a mix of qualitative and quantitative information. The questionnaires were sent by email to the EU Ecolabel awarded mountain huts after having introduced the purpose of the survey by phone to the contact person for the EU Ecolabelled service. In some cases completed questionnaires were sent back, whereas other questionnaires were partially or fully completed via a phone interview. The questionnaire respondents were mainly the hut-keepers of the mountain huts that possessed EU Ecolabel in May 2012 (Table 24.1) as reported by the National Register updated by ISPRA for the Italian Competent Body 'Comitato Ecolabel-Ecoaudit'.

24.3 Results and Discussion

Among the existing voluntary policy initiatives that an organization may implement to contribute directly and indirectly to the efficient use of (natural) resources and to the environmental protection, formal certification schemes concerning products, services or the organization itself are the ones which may guarantee a defined framework for an official recognition of company's environmental commitment to continuous improvement. However, it is not always so obvious to the public the distinction between process- or performance-based certification programs. Process-based certification focuses on implementation and maintenance of environmental policies, management and operational procedures through a company-wide approach. In that case the certification is attained once that all the requirements established by the reference standard (e.g. ISO 14001 for environmental management) have been met and the compliance is verified by an independent body. Instead, the performance-based certification focuses on specific environmental (and sometimes also economic and social) criteria for a product or service that must be fulfilled and independent third-party verified to receive the statement of compliance.

The performance-based certification is the scheme implied by the third party eco-label, also known as Type I environmental labeling, run by independent organizations (Ecolabelling body), for which the compliance of pre-set criteria, established on the basis of life cycle inventory and sometimes impact assessment, is required. EU Ecolabel is a typical example of the Type I scheme as it is intended to ensure both a high level of environmental performance of the company's product or service with regard to certain aspects, and offer the corresponding information to the consumer. Among the voluntary schemes the EU Ecolabel is commonly recognized as a commitment to sustainability mainly aimed at excellence in the environmental management. In particular, for tourist accommodation service it has been established that beside the fulfillment of a certain number of mandatory criteria (in total 29 if applicable), also a set of optional criteria (in total 61), defined with a point system, has to be satisfied by the applicant for a minimum of twenty points. According to the life cycle approach, the ecological performance criteria are referred to the before-service phase (purchasing), operational phase (service provision), and the after-service phase (waste). The general aim is intended to: (a) limit energy and water consumption, (b) limit waste production, (c) favor the use of renewable resources and substances less hazardous to the environment, and (d) promote environmental communication and education.

The data analysis here reported was intended to highlight some peculiar aspects concerning the perception of the European scheme by the Italian hut-keepers interviewed. For this purpose a questionnaire was designed to gather a mix of qualitative and quantitative information. The questionnaire was divided into five sections as follows:

- Questions from 1 to 6 were intended to gather data about the reasons that led to the decision to apply for the EU Ecolabel and details on the certification project;
- Questions from 7 to 10 were aimed at identifying the perceived advantages derived from the EU Ecolabel and the difficulties encountered in ecolabelling implementation;
- Questions from 11 to 16 concerned the maintenance process of the certification;
- Questions from 17 to 22 referred to the external communication practices and the environmental awareness/knowledge, particularly of the EU Ecolabel, among the guests;
- Questions from 23 to 28 were aimed to obtain details about the best practices adopted for the environmental management, the environmental education and communication to the guest and other stakeholders.

Fig. 24.2 Participation to other certification schemes

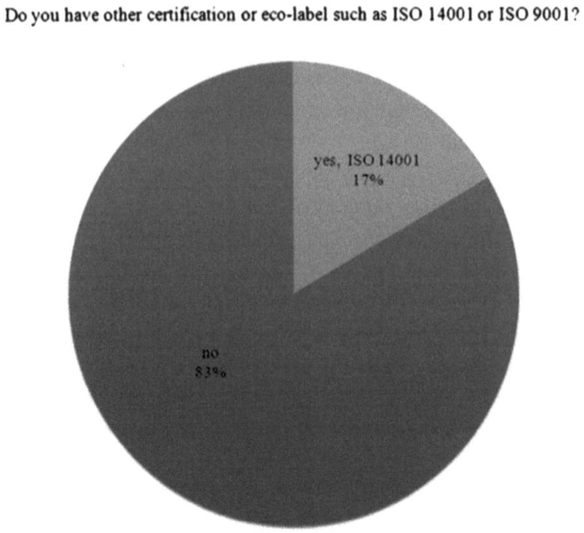

The main findings of this analysis have been represented graphically in Figs. 24.2, 24.3, 24.4 and 24.5 and concern specifically (a) the participation to other certification schemes, (b) the perception of the advantages derived from the EU Ecolabel participation, (c) the willingness of the hut-keepers to recommend the EU Ecolabel to others, and (d) the customer awareness about the EU Ecolabel meaning.

As regards the participation in other certification schemes, process-based or performance-based, the data show that only in one case a mountain hut has declared the acquisition of the ISO 14001 certification for the environmental management system (Fig. 24.2). Even for the service provided by a mountain hut operating procedures and instructions may be implemented for a more sustainable impact on the environment (Beltramo and Cuzzolin 2001). The implementation of an environmental management will represent also in this case a key factor for on-going performance improvement. Though eco-labels represent a valuable tool for sustainable tourism, they will likely be most effective if used along with other environmental management tools, as part of an integrated strategy. An organization with an environmental management system (EMS) certified ISO 14001 or with the EMAS registration can benefit from using the Ecolabel criteria in its environmental policy with reference to the product performance targets. An organization with Ecolabel instead would benefit from the adoption of an EMS to maintain its compliance with the required Ecolabel criteria. As established by the European Regulation, the product group criteria are periodically subject to possible revision in order to ensure higher environmental standards.

As to the advantages of European Flower perceived by the participants on their own business activities, the graph reported in Fig. 24.3 shows that the two most important benefits are mainly associated to an improved image and a reduction of the environmental impact. EU Ecolabel is often able to produce an improvement in

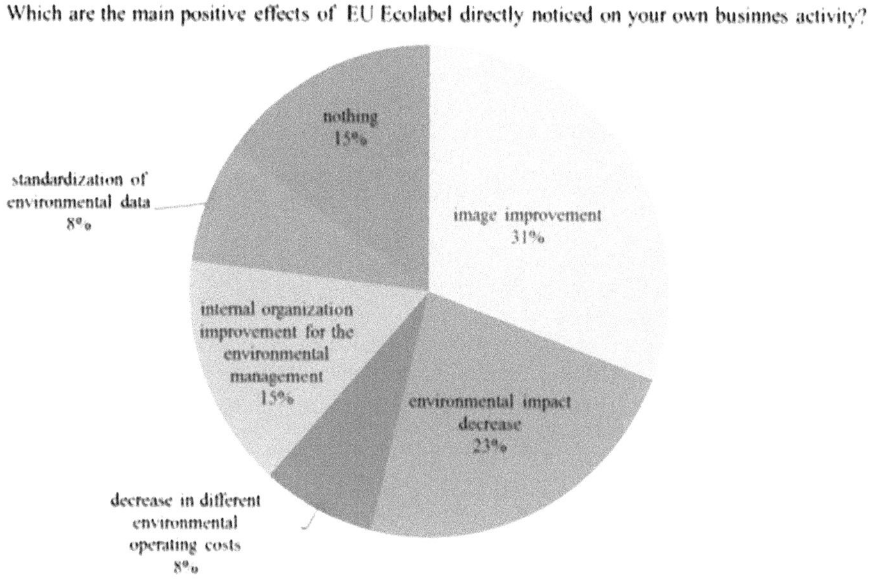

Fig. 24.3 The perceived advantages of the EU Ecolabel

Fig. 24.4 Willingness to recommend the EU Ecolabel

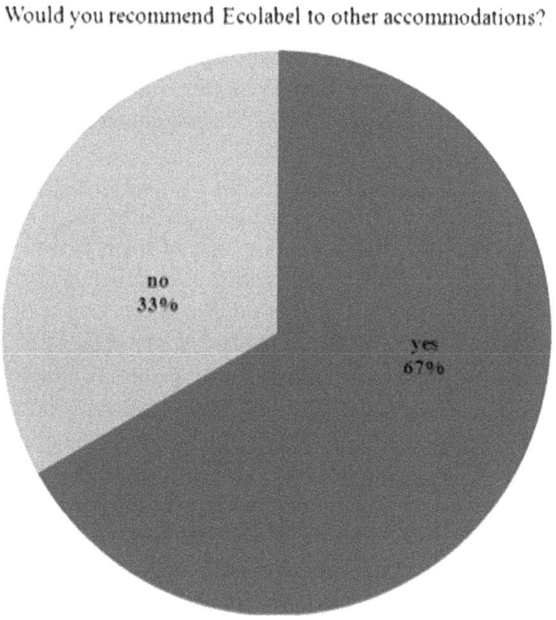

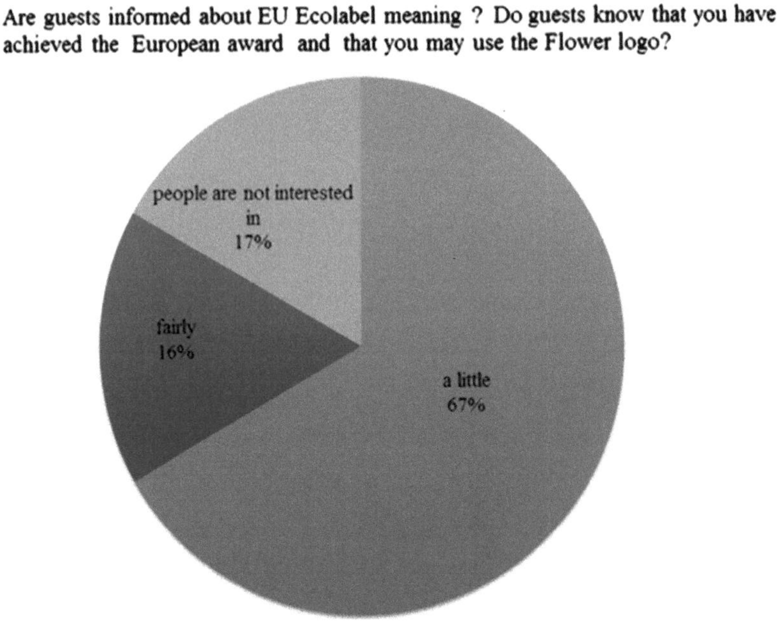

Fig. 24.5 Customer awareness of the EU Ecolabel meaning

the environmental performance and consequently a lower impact. Other benefits recognized by the respondents are an improvement of the internal organization, a more eco-efficient operational management and the standardization of the environmental data.

Regarding the willingness of the hut-keepers to recommend the EU Ecolabel to others (Fig. 24.4), the answers given by the mountain-hut keepers were mainly positive with two exceptions; one of them is now corresponding to a participant dropping out. In general, certified companies willing to recommend a certification scheme are those whose expectations had been met or those who believed the scheme to be cost-effective. Here, it should also be mentioned that in the last two years the number of Italian EU Ecolabel awarded mountain huts is decreased by 50 %. A negative trend in Italy that would confirm some problems lied to the procedures for the implementation and maintenance of the Community award. It should be mentioned that the EU Ecolabel for the mountain huts is still solely an Italian phenomenon and now mainly delimited to the Region Piemonte.

As far as the awareness and knowledge about the European eco-label's meaning is considered, also among the mountain-hut guests it has been pointed out that information about the meaning of the EU Ecolabel and the environmental award recognized to the accommodation service is relatively low (Fig. 24.5). As reported in the Eurobarometer on 'European's attitude towards the issue of sustainable consumption and production', the awareness of the EU Ecolabel and its Flower logo in Italy is one of the lowest along with UK and Sweden in Europe (The European Commission 2009).

Therefore, not even in the mountain context the visibility of the European logo for green accommodation may be considered satisfying to attract new guests and hence new applicants for the Ecolabel award. Generally speaking, ecolabelling would be most attractive for the service provider if it would be a customer demand/interest or if at least the efforts sustained to obtain the environmental award were recognized by them.

24.4 Conclusions

On the basis of the statistics the European Flower is in full bloom. Between 2000 and 2010 the number of licenses was multiplied by a factor of more than 20, and this results was due to the tourism accommodation service sector particularly, where the number of licenses are still increasing rapidly. In particular, thanks to a lot of promotion and communication initiatives carried out at a regional level, Italy is the country with the major percentage of Ecolabel holders (31 % of the European total number) and the greatest number of holders in the tourist accommodation product group.

Among the accommodation services that may be Ecolabel awarded we find also the mountain huts, even if at the moment it is an all-Italian phenomenon. In no other country we may find mountain huts awarded with the European eco-label.

By analyzing the preliminary results here reported, to be treated with caution given the relative low number of EU Ecolabelled mountain huts now existing, the main conclusion is that the experience of some past and present holders is not fully positive. The number of the Ecolabel dropouts for this specific type of accommodation service in the last years could not have differently been explained. In particular, the interviews showed that even if positive effects are recognized by the respondents, mainly in terms of image improvement and environmental impact reduction, the promotion initiatives of the EU Ecolabel are not considered so effective since generally the guests know little about the European Flower and its meaning. Moreover, there are some subjects among the respondents that are not willing to recommend the EU Ecolabel to other accommodation.

The negative features of the experience of participation in the European Ecolabel here identified undoubtedly call for further investigations. A follow-up of the study would permit to verify the perception of the European scheme by other accommodation types with a data stratification performed according to factors not included in this analysis.

References

Beltramo R, Cuzzolin B (2001) Manual for the implementation of an environmental management system for mountain refuges. Editions L'Eubage, Aosta

Buckley R (2002) Tourism ecolabels. Ann Tourism Res 29(1):183–208

Chen A, Peng N (2012) Green hotel knowledge and tourists' staying behaviour. Ann Tourism Res 39(4):2211–2216

Commission of the European Communities (2009) Commission Decision of 2009/578/EC of 9 July 2009 establishing the ecological criteria for the award of the Community eco-label for tourist accommodation service. OJ L 198/57, 30.7.2009

ISPRA Ecolabel EU Servizi Certificati (2012) ISPRA, Rome. http://www.isprambiente.gov.it/it/certificazioni/ecolabel-ue/servizi-certificati. Accessed 3 May 2012

ISPRA (2009a) Tourism accommodation and camp site services EU eco-label award scheme. Final activity report to criteria revision camp site service and tourism accommodation service. ISPRA, Rome

ISPRA (2009b) User manual for the eco-label for tourist accommodation service. ISPRA, Rome

ISPRA (2012) L'Ecolabel dell'Unione europea per le strutture ricettive e per i campeggi. Un marchio di eccellenza ambientale. Tiburtini s.r.l., Rome

European Commission (2009) Europeans' attitudes towards the issue of sustainable consumption and production. Analytical Report. Flasheurobarometer 256, July 2009

European Commission (2011) EU Ecolabel Work Plan for 2011–2015

European Parliament and the Council of European Community (2010) Regulation (EC) No 66/2010 of 25 November 2009 on the EU Ecolabel. OJ L 27/1, 30.1.2010

Eurostat (2011) Sustainable development in the European Union. 2011 monitoring report of the EU sustainable development strategy. European Union, Luxembourg

United Nations World Tourism Organization (UNWTO) (2002) Voluntary Initiatives for Sustainable Tourism. UNWTO, Madrid

United Nations World Tourism Organization (UNWTO) (2003) Recommendations to Governments for Supporting and/or Establishing National Certification Systems for Sustainable Tourism. UNWTO, Madrid

Chapter 25
CAP and Sustainability. Subsidies and Biodiversity in Six "Objective-1" Italian Regions

Alba Marino, Guido Signorino and Elisa Gatto

Abstract Biodiversity is a mean and a goal of sustainability and should be preserved and encouraged. In order to protect the agricultural sector in Europe and to stabilize farmers' revenues and incomes, starting from the 1950s the Common Agricultural Policy (CAP) established a subsidies regime. Typically, policy interventions may produce two undesirable outcomes: (1) distortion of farmers' preferences, averting agricultural sector from economic efficiency; (2) production concentration towards protected cultivations, so reducing biodiversity and hindering the sustainability of agricultural activity. Under this perspective, the sustainability of agricultural policies should be evaluated in terms of their effect on agricultural biodiversity, and a trade-off between subsidies and biodiversity in agriculture is expected. This chapter investigates the relationship between subsidies and biodiversity in six Italian "Objective-1" regions, finding a negative and quite stable correlation between these variables. Recent CAP reforms aimed at reducing market distortions through the adoption of "decoupled" measures, that supported agricultural activity regardless of the type of crop produced are expected to bring positive effects on agricultural biodiversity and sustainability.

Keywords Common Agricultural Policy · Biodiversity · Sustainability · Subsidies · Agricultural risk

25.1 Sustainability and Biodiversity

During 1960s and 1970s new ecologist and social movements have had great influence on innovative cultural perspective leading to centrality of the concept of sustainability to underline environmental protection. Since then, stakeholders' attention has being interested in sustainability and development, especially in managing natural resources, soil erosion, climate changes, water quality and quantity, poverty or social exclusion. The concept of sustainability, used not only in ecological and scientific literature, but

G. Signorino (✉) · A. Marino · E. Gatto
Department of Economics, Business, Environment and Quantitative Methods, University of Messina, Piazza S. Pugliatti 1, 98122 Messina, Italy
e-mail: signorin@unime.it

R. Salomone, G. Saija (eds.), *Pathways to Environmental Sustainability*,
DOI 10.1007/978-3-319-03826-1_25,

also in economic and social context, was defined by the Brundtland Report (United Nations) on March 20, 1987: "*it is development that meets the needs of the present without compromising the ability of future generations to meet their own needs*".

According to the Convention on Biological Diversity (CBD), biodiversity "[…] means the variability among living organisms from all sources including, inter alia, terrestrial, marine and other aquatic ecosystems and the ecological complexes of which they are part; this includes diversity within species, between species and of ecosystems"[1].

Biodiversity is a composite phenomenon that produces heterogeneous effects in environmental field, such as economic or social ones. For this purpose, it is necessary to encourage a multidisciplinary approach and new types of communication and cooperation, e.g. among agriculturalists, ecologists, and economists to identify and establish adequate assessment strategies (Robertson and Swinton 2005).

Biological diversity not only improves crop production processes, but provides important ecosystem and environmental services[2] necessary for the sustainability of human development and for influencing a higher resilience[3]. Hence, the complexity of the combined ecological and social functions of agrobiodiversity has encouraged several authors to deal with these issues (e.g. Altieri 1999; Jackson et al. 2007).

In the Millennium Ecosystem Assessment (MEA 2005), biodiversity was viewed as an important coping strategy against agricultural risks in an uncertain future, but with the current state of knowledge, this may be viewed as *received wisdom* rather than substantiated proof of process (Wood and Lennè 2005).

Although there are several theoretical papers about the insurance value of agrobiodiversity, few empirical studies has been realized (e.g., Qualset et al. 1995; Smale et al. 1998; Di Falco and Chavas 2004; 2009, Figge 2004; Cafiero et al. 2007; Barkley and Peterson 2008).

25.2 The Evolution of Common Agricultural Policy (CAP)

Agriculture raised significant importance in EU policy schemes[4] and drawn attention of policy makers and researchers all over the world. Furthermore biological nature and multifunctional vocation of the agricultural sector led governments—and other stakeholders—to assert that farming is riskier than other enterprises and to underline its function about resource conservation and general environmental aware-

[1] Art. 2, Convention on Biological Diversity. United Nations Conference on Environment and Development, Rio De Janeiro, 1992.

[2] For example recycling of nutrients, control of local microclimate and hydrological processes, pest regulation and depuration from noxious chemicals.

[3] The capacity of a system to absorb disturbance and reorganize while undergoing change so as to still retain essentially the same function, structure, identity, and feedbacks.

[4] "*Sustainable agriculture is not a luxury but a necessity. It concerns us all, whether we live in Europe, or on any other place on Earth*" (Dacian Cioloş, European Commissioner for Agriculture and Rural Development).

ness[5]. For these reasons, an international debate on the role and performance of risk management instruments has been encouraged and public support to agriculture justified.

As agricultural economic literature recognized, crop-diversity[6] represents an efficient strategy to cope with risk in agriculture and to reduce farmers' exposure and vulnerability to risk. Some empirical studies emphasized the existence of a negative long-run relationship between high returns and yields or farm incomes variability (Smale et al. 1998; Di Falco and Chavas 2007). Di Falco and Chavas (2009) confirmed that genetic diversity could support agricultural productivity and help to reduce the risk of crop failure, especially in degraded land. Referring to developed economies, Di Falco and Perrings (2003) built up a theoretical and empirical investigation on a diversity-productivity and diversity-stability hypothesis. The model was tested on data from Southern Italy, finding a positive relationship of crop diversity to mean income and a negative one to the variance of income. All these studies could be connected as an application of some kind of "crop-folio theory"[7]: the study of decision strategies under risks has a long tradition as well as many applications in agricultural choices[8].

For many strategic, political, economic and social reasons, governments used to intervene in the agricultural sector in an attempt to regulate the production and trade. The Common Agricultural Policy was firstly identified in the Founding Treaty of the European Community (1957) which defined its purpose and aims "to increase agricultural productivity, to ensure thereby a fair standard of living for the agricultural population, to stabilize markets, to guarantee regular supplies, and to ensure reasonable prices in supplies to consumers". The first CAP regulation was characterized by a *double core* oriented to price and structural policies: price guarantees, protectionist trade policy (such as variable import levies or export subsidies), but also support farm modernization and professional qualification.

In 1992 a reform of the Common Agricultural Policy had become inevitable. The MacSharry Reform was characterized by gradual market re-orientation of production decisions through reduction of intervention prices and introduction of partially decoupled payments[9] in order to limit supply surpluses and EU expenditure.

This action represented a fundamental shift in policy-making process, underling the difference between "old CAP" and the "new" one. Although MacSharry Reform did not reduce inequity in the balance between structural policy and market

[5] In this sense, for example, European organizations have drawn a long-run biodiversity strategy, taking into account the important role of agricultural activities (see EC Commission 2011).

[6] Crop-diversity could be considered as a subcategory of biodiversity in agroecosystem.

[7] Figge (2004) referred to studies about risk-reducing effect and economic value of genetic diversification in agricultural decisions as "portfolio effect applications" or "bio-folio analysis". Similarly we refers to farmers' diversification through different cultivations as "crop-folio decisions".

[8] There are some interesting summaries on portfolio applications to biodiversity (e.g. Figge 2004; Barkley and Peterson 2008).

[9] Payments set on cultivated lands and differentiated across products. They were no more linked to the quantity produced, but based on variety of crops produced in order to avoid increasing production without additional market demand.

policy, the proposal allowed price to partly regains its function of market signal for farmers' cultivation decisions.

The CAP was finally reorganized as a two-track system introducing a "first pillar" (market policy and direct payments) and a "second pillar" (structural and rural development policy). Afterwards, a *Mid-Term Review*[10] promoted a totally decoupled support system and a progressive alignment of domestic prices towards global market. Over the course of forty years the Common Agricultural Policy had moved from *subsidy-oriented* production, where farmers' decisions were dictated by the entity of support that could be obtained, to a *market-oriented* production, characterized by central role of market conditions in taking decisions.

25.3 The Biodiversity-Subsidy Trade-Off

As price instability is one of the most important component of economic risk in agriculture, subsidies on specific cultivations may alter market risk conditions and distort farmers' preferences. It is possible to verify, as a result of that observation, a trade-off between subsidies and biodiversity (Gatto and Signorino 2011).

A particular analysis regarding a significant sample of Norwegian farms[11] showed that subsidy scheme or market conditions could be more important than farmers' risk-aversion on portfolio choice in agriculture (Lien and Hardaker 2001). According to these results, financial assistance could influence farmers' decisions more than innate risk aversion elements. However, risk preferences may play a pivotal role in determining crop biodiversity. In fact, if the coexistence of similar types of crops which react differently to environmental or market risks is a risk-reducing strategy, risk-averse farmers may choose to control their risk exposure through diversification (Meng 1997; Heal 2000; Di Falco and Perrings 2003, 2005).

As we have already mentioned, risk in agriculture is a complex topic, given that farmers face more sources of risk. Besides the "market risk" that typically characterizes entrepreneurial activity in a competitive environment (price variability), a "natural risk" has to be considered, due to uncertainty about aspects like meteorology and/or pathogenic agents that may damage harvests.

In the presence of subsidies aimed at protecting specific cultivations, farmers may activate two different (not necessarily alternative) strategies in order to cope with risk. The first one is to increase the quota of protected cultivations within their total production (so concentrating their supply in those products that reduce their market risk); the second one is to increase the diversity of their production (so increasing soil productivity and reducing the "natural risk", especially as far as the exposure of specific cultivation to pathogenic factors is concerned). To the extent

[10] The Mid-Term Review was approved in 2003 and it is also known as Fischler Reform.

[11] Norwegian agriculture is characterized by relatively large subsidies compared to other countries. It is justified on particular national conditions of natural landscapes as well as social and economics context.

that sustainability is associated with biodiversity, a higher degree of agricultural diversity stands for a more "sustainable" production.

Hence, subsidies to specific cultivations may have a double negative outcome on economic performance and on sustainability: on the first side, they influence farmers' preferences averting their choices from market efficiency; on the second one, they reduce biodiversity in agriculture.

As a result, within certain limits, the two strategies are alternative means of risk reduction, and increased use of subsidies leads to lower crop biodiversity, requiring a greater production concentration in favor of subsidized crops. As a result, at least as long as the benefit system is based mainly on "coupled" transfers (i.e. related to specific crops), subsidies and biodiversity are substitutes in farmers' risk protection strategies, and a trade-off between the two variables is expected to be found analyzing firms' behavior.

In fact, subsidies that are directed at protecting specific cultivations promote specialization strategies that increases (both in absolute and in relative terms) the production of the protected species, contrasting biodiversity (or the diversification of farmers' productive portfolio). As a consequence, turning to "coupled" subsidies should reduce agricultural diversity of production.

25.4 The Empirical Results

As seen above, most of the applied research estimates the role of biodiversity and subsidies on farms' economic performance by implementing standard production functions that include a diversity and a subsidy (or a policy) variable within the regressors set.

The scope of this empirical section is to directly test the stated relationship between biodiversity and subsidies under the period of "coupled" subsidies that characterized CAP up to the first half of 2000s, in order to infer the existence of the expected trade-off.

Based on the RICA (Rete Informazione Contabile Agricola) database, the relationship between subsidies and biodiversity in agriculture has been examined by running a correlation analysis and a regression exercise on the data of six Regions in the Italian Mezzogiorno ("Objective-1" areas of the European Community). While the correlation analysis is carried out on average values at a regional level, as will be explained below, the regression technique is applied in order to further investigate on the nature of the variables' relationship at an individual level, as a validation of the correlation results.

Available data cover the period 1995–2005[12]. RICA database is provided by the Italian National Institute of Agricultural Economics (INEA); information is collected according to the standard of the European Commission "Farm Account-

[12] The Regions of Sicily and Sardinia have been excluded because data were only available starting from the year 2000.

ing Data Network" (FADN), whose RICA represents the Italian component. RICA collects annual information from more than 15,000 Italian agricultural firms, preserving a representative criterion at a regional level.

Both the Biodiversity index (B) and the Subsidy variable were obtained from the RICA database. The first variable was calculated at a firm level using the Simpson spatial diversity index: $D = \sum_{i=1}^{s} \left(\frac{s_i}{s_{tot}} \right)^2$, where s_i is surface devoted to the cultivation of the *i-th* production, and s_{tot} is the total cultivated extension of the firm. Simpson's spatial diversity index D is a concentration index, as its range is $0 < D \leq 1$ and its value is obviously 1 in the case of complete specialization, when only one species is cultivated and $s_1 = s_{tot}$. A more direct "biodiversity" index is simply obtained subtracting to unity the Simpson's spatial diversity statistics:

$$B = 1 - D = 1 - \sum_{i=1}^{s} \left(\frac{s_i}{s_{tot}} \right)^2$$

The Subsidy variable is the total amount of transfers that individual farms received by EU.

As it is well known, the Pearson's coefficient (r_{xy}) detects the existence of a systematic (positive or negative) correlation between two variables (x, y), being obtained by dividing the covariance of the two variables by the product of their standard deviations: $r_{xy} = \sigma_{xy} / (\sigma_x \, \sigma_y)$. The correlation coefficient ranges between -1 and $+1$, and the closer its value is to unity (in absolute terms), the stronger is the statistical linkage (the correlation) between the two variables. According to the hypothesis of a trade-off between Biodiversity (B) and Subsidies (S), the expected sign of the correlation is negative.

The correlation analysis was conducted by estimating the average annual regional values of both biodiversity and subsidies for the period and calculating the Pearson's coefficient for the two variables at a regional level. Given that the correlation exercise is run on a limited number of observations, a non-parametric correlation statistics (the Spearman's rank correlation coefficient ρ) has also been estimated in order to corroborate the results. The Table 25.1 shows the results of the estimation.

As expected, $r_{B,\,S} < 0$ in all regions and the coefficient (excluding the case of Calabria) shows values that are close to or significantly higher than $|0.5|$. This reveals that, on average, a higher level of subsidies is associated with a lower value of regional biodiversity in agriculture. Further, with the exception of Calabria, an important negative correlation has been detected also by using Spearman's non-parametric ranking correlation ρ.

The regression analysis was aimed at evaluating the strength and significance over time of the relationship between the two variables. Under the assumption that in absence of subsidies agricultural firms would structure a more differentiated portfolio of cultivations in order to reduce the "natural" risk of production and increase

Table 25.1 The Pearson's and Spearman's coefficients between biodiversity and subsidies

Region	Pearson's r	Spearman's ρ
Abruzzo	−0.682	−0.800
Molise	−0.555	−0.510
Campania	−0.414	−0.691
Calabria	−0.007	0.009
Puglia	−0.712	−0.697
Basilicata	−0.560	−0.790

firm returns, firm-level biodiversity has been regressed versus subsidy, according to the model: $b=\beta\ (s)$, where b=farm's biodiversity (or Cultivation Diversity) and s=farm's subsidies/gross saleable production.

The regression model: $b_j=\alpha_0\ s_j+cost+\varepsilon$ has been estimated with STATA 11.0 using OLS on a regional basis year by year for the period 1995–2005 (see Table 25.2). As highlighted above, this is not a complete model for agricultural biodiversity; rather, our aim with this exercise is to control for the sign, the statistical significance and the confidence interval of the estimated coefficients. In other words, given that at an aggregate level increases in the amount of subsidies perceived by farms in each region are negatively associated with spatial diversity, we expect, in general, to find $\alpha_0<0$, and we want to test the significance of the coefficients and the sign of both the upper and the lower bounds of their confidence interval.

The results of the estimations are reported in Tables 25.2 (coefficients and their p-values) and 25.3 (confidence intervals). Empty cells in Tables 25.2 and 25.3 are due to shortage of information: we only considered years with more than 100 observations.

In Table 25.2 values in bold are associated with a confidence interval entirely lying in the negative (or positive) set; In general, in such a case the coefficient is also significant (p-value<.05), as in Table 25.3, where bold is assigned to the confidence intervals associated with a significant coefficient. Apart from the case of Campania, until 2000–2001 the sign of the coefficient is high and in negative most cases. Both the sign and the significance of the parameter become less clear in the period 2001–2005, when the CAP reform, introducing "delinked" subsidies, started. As a consequence, results show that in the period of "linked" subsidies a trade-off characterized the biodiversity-subsidies relationship.

25.5 Conclusion

This chapter aimed at evaluating the sustainability of Common Agricultural Policy. The trade-off between biodiversity and subsidies of Southern Italian firms in facing risk in agriculture during the period 1995–2005 was scrutinized with both a correlation and a regression analysis. In accordance with the portfolio (or bio-folio) approach, we found that, both the correlation and the regression coefficients as-

Table 25.2 Regression analysis—estimated coefficients and *p*-values

		Abruzzo	Molise	Campania	Calabria	Puglia	Basilicata
1995	Coeff.	**−0.846**					**−0.751**
	P-value	**0.000**					**0.000**
1996	Coeff.	**−0.476**		**−0.0001**	**−1.070**	**−0.001**	−0.0001
	P-value	**0.020**		**0.030**	**0.000**	**0.004**	**0.244**
1997	Coeff.	**−0.793**		**0.0001**	**−0.0002**		**0.0001**
	P-value	**0.000**		**0.001**	**0.000**		**0.000**
1998	Coeff.	**−0.616**	**−0.190**	**0.982**	**−1.316**		**−0.831**
	P-value	**0.002**	**0.265**	**0.002**	**0.000**		**0.000**
1999	Coeff.	**−0.641**	**−0.586**		**−1.322**		**−0.792**
	P-value	**0.015**	**0.000**		**0.000**		**0.000**
2000	Coeff.	**−0.246**	**−0.079**	**0.185**	**0.563**	**−0.205**	**−0.205**
	P-value	**0.001**	**0.003**	**0.000**	**0.000**	**0.001**	**0.001**
2001	Coeff.	−0.037	0.010	**0.136**	**−0.274**	**−0.127**	**−0.829**
	P-value	0.130	0.849	**0.007**	**0.000**	**0.001**	**0.000**
2002	Coeff.	**0.131**	**−1.560**	**0.089**	**−0.374**	5.296	
	P-value	**0.001**	**0.028**	**0.000**	**0.000**	0.215	
2003	Coeff.	0.047	**−0.074**	**0.345**	−0.118	**−0.294**	−0.351
	P-value	0.086	**0.018**	**0.000**	0.050	**0.000**	0.000
2004	Coeff.	**0.107**	−0.050	**0.517**	**−0.208**	**−0.212**	−0.339
	P-value	**0.000**	0.318	**0.000**	**0.004**	**0.000**	0.000
2005	Coeff.	−0.025	0.123	**−0.528**	3.478	−5.820	−0.201
	P-value	0.303	0.083	**0.000**	0.969	0.917	0.000

Level of significance of 95 %

sume in general the expected negative signs, thereby evidencing that the increases of diversity are associated with subsidy decreases. Yearly regression analysis based on data from single firms in each region shows that in the final part of the period, when a delinked subsidization has been gradually introduced, this trade-off became less general and strong.

The conclusion is that the two variables are substitutes in the firms' strategies of risk protection, so that turning to subsidies reduces biodiversity in agriculture, even though diversity is not explained by subsidies. In the light of the "portfolio approach" that we have briefly introduced, this may imply that subsidies influence the "gross" composition of the productive portfolio (differentiating the quotas of "risky" and "non-risky" species), while the total differentiation of productions responds to other variables (e.g., height, soil fertility, meteorological information, market risk for individual cultivations, etc.).

Since the purpose of this chapter was to investigate the relationship between subsidy and productive diversity under the CPA regime, we have been able to show that until 2006 the "coupled" subsidy regime that characterized agricultural policy in the EU gave rise to a trade-off between policy interventions and productive diversity. Recent CAP reforms which introduced the instrument of "decoupled" subsidies are expected to relax this trade-off and to increase the sustainability of agricultural policies.

Table 25.3 Regression analysis—confidence intervals

		Abruzzo	Molise	Campania	Calabria	Puglia	Basilicata
1995	Lower	**−1.093**					**−1.042**
	Upper	**−0.599**					**−0.543**
1996	Lower	**−0.877**		**−0.0002**	**−1.527**	**−0.001**	−0.0003
	Upper	**−0.074**		**−0.0001**	**−0.613**	**−0.0001**	0.0001
1997	Lower	**−1.189**		**0.0001**	**−0.0003**		**−0.00008**
	Upper	**−0.397**		**0.0002**	**−0.0001**		**−0.00002**
1998	Lower	**−1.012**	−0.524	**0.359**	**−1.835**		**−1.085**
	Upper	**−0.220**	−0.144	**1.605**	**−0.780**		**−0.577**
1999	Lower	**−1.153**	**−0.878**		**−1.776**		**−1.042**
	Upper	**−0.128**	**−0.292**		**−0.867**		**−0.543**
2000	Lower	**−0.387**	**−0.130**	**0.100**	**−0.668**	**−0.320**	**−0.320**
	Upper	**−0.106**	**−0.027**	**0.271**	**−0.458**	**−0.090**	**−0.090**
2001	Lower	−0.086	−0.098	**0.038**	**−0.388**	**−0.199**	**−1.186**
	Upper	0.011	0.119	**0.234**	**−0.159**	**−0.056**	**−0.471**
2002	Lower	**0.050**	**−0.295**	**0.050**	**−0.506**	−3.087	
	Upper	**0.211**	**−0.017**	**0.128**	**−0.242**	13.678	
2003	Lower	−0.007	**−0.183**	**0.229**	−0.237	**−0.362**	**−0.432**
	Upper	0.100	**0.035**	**0.462**	0.000	**−0.226**	**−0.270**
2004	Lower	**0.026**	−0.148	**0.386**	−0.349	**−0.276**	**−0.415**
	Upper	**0.187**	0.048	**0.647**	−0.068	**−0.148**	**−0.264**
2005	Lower	−0.072	−0.016	**−0.654**	−128.139	−115.128	**−0.260**
	Upper	0.023	0.263	**−0.402**	135.095	104.470	**−0.142**

Level of significance of 95 %

References

Altieri MA (1999) The ecological role of biodiversity in agroecosystem. Agric Ecosyst Environ 74:19–31

Barkley A, Peterson HH (2008) Wheat variety selection: an application of portfolio theory to improve returns. Proceedings of the NCCC-134 Conference on Applied Commodity Price Analysis, Forecasting and Market Risk Management. St. Louis, MO

Cafiero C, Capitanio F, Cioffi A, Coppola A (2007) La gestione del rischio nelle imprese agricole tra strumenti privati e intervento pubblico. Agriregioni Europa 3(8). http://www.agriregionieuropa.univpm.it/dettart.php?id_articolo=199. Accessed 3 Dec 2012

Di Falco S, Chavas J-P (2004) Crop biodiversity, farm productivity and the management of environmental risk. Eur Rev Agric Econ 33(3):289–314

Di Falco S, Chavas JP (2007) Farmer management of production risk on degraded lands: the role of wheat variety diversity in the Tigray region, Ethiopia. Agric Econ36 (2): 147–156

Di Falco S, Chavas J-P (2009) On crop biodiversity, risk exposure and food security in the highlands of Ethiopia. Am J Agric Econ 91(3):599–611

Di Falco S, Perrings C (2003) Crop genetic diversity, productivity and stability of agroecosystems. A theoretical and empirical investigation. Scott J Polit Econ 50(2):207–216

Di Falco S, Perrings C (2005) Crop biodiversity, risk management and the implications of agricultural assistance. Ecol Econ 55(4):459–466

EC Commission (2011) Our life insurance, our natural capital: an EU biodiversity strategy to 2020, COM (2011) 244 final

Figge F (2004) Bio-folio: applying portfolio theory to biodiversity. Biodivers Conserv 13:822–849

Gatto E, Signorino G (2011) Long-run relationship between crop-biodiversity and cereal production under the CAP reform: evidence from Italian regions. Paper presented at ERSA Conference, Barcelona

Heal G (2000) Nature and the marketplace: capturing the value of ecosystem services. Island Press, New York

Jackson LE, Pascual U, Hodgkin T (2007) Utilizing and conserving agrobiodiversity in agricultural landscapes. Agric Ecosyst Environ 121(3):196–210

Lien G, Hardaker JB (2001) Whole-farm planning under uncertainty: impacts of subsidy scheme and utility function on portfolio choice in Norwegian agriculture. Eur Rev Agric Econ 28(1):17–36

Meng ECH, Smale M, Bellon MR, Grimanelli D (1998) Definition and measurement of crop diversity for economic analysis. In: Smale M Farmers, gene banks, and crop breeding. Kluwer, Boston, pp 19–31

MEA (2005). Synthesis Report. Island Press, Washington, DC

Qualset CO, McGuire PE, Warburton ML (1995) Agro-biodiversity: key to agricultural productivity. Calif Agric 49(6):45–49

Robertson GP, Swinton SM (2005) Reconciling agricultural productivity and environmental integrity: a grand challenge for agriculture. Front Ecol Environ 3:38–46

Smale M, Hartell J, Heisey PW, Senaue B (1998) The contribution of genetic resources and diversity to wheat production in the Punjab of Pakistan. Am J Agr Econ 80(3):482–493

Wood D, Lennè JM (2005) 'Received wisdom' in agricultural land use policy: 10 years on from Rio. Land Use Policy 2(2):75–93

MIX
Papier aus verantwortungsvollen Quellen
Paper from responsible sources
FSC® C105338

If you have any concerns about our products,
you can contact us on
ProductSafety@springernature.com

In case Publisher is established outside the EU,
the EU authorized representative is:
Springer Nature Customer Service Center GmbH
Europaplatz 3, 69115 Heidelberg, Germany

Printed by Libri Plureos GmbH
in Hamburg, Germany